STATE OF THE CLIMATE 2017

Ole Humlum

ISBN 978-0-9931190-2-6
Printed by Createspace.
Cover image ©Shutterstock.

Published by:
The Global Warming Policy Foundation
55 Tufton Street
London SW1P 3QL

www.thegwpf.org

Contents

About the author

Ole Humlum is former Professor of Physical Geography at the University Centre in Svalbard, Norway, and Emeritus Professor of Physical Geography, University of Oslo, Norway.

Executive summary

1. It is likely that 2017 was one of the warmest years, according to temperature records from the instrumental period (since about 1850). However, it was cooler than 2016.

2. At the end of 2017 the average global air temperature was dropping back towards the level characterising the years before the strong 2015–16 oceanographic El Niño episode. This underscores that the global surface temperature peak of 2015–16 was caused mainly by this Pacific oceanographic phenomenon. It also suggests that what has been termed 'the temperature pause', 'hiatus', or similar terms, may continue in years to come.

3. There still appears to be a systematic difference between average global air temperatures estimated by surface stations and by satellites. Especially since 2003, the average global temperature estimate derived from surface stations has steadily drifted away from the satellite-based estimate in a warm direction.

4. The temperature variations recorded in the lowermost troposphere are generally reflected at higher altitudes also, and the overall temperature 'pause' since about year 2002 is recorded at all altitudes, including the tropopause and into the stratosphere above. In the stratosphere, however, the temperature pause had already begu by around 1995; that is, 5–7 years before a similar temperature pause began in the lower troposphere near the planet's surface. The stratospheric temperature pause has now existed without interruption for about 23 years, with no explanation offered by the climatological community.

5. The recent 2015–16 oceanographic El Niño episode is among the strongest since the beginning of the record in 1950. Considering the entire record, however, recent variations between El Niño and La Niña episodes are not unusual. In late 2017, a new cold La Niña episode appears to have begun to emerge. If this continues, it will influence global temperatures in 2018.

6. Since 2004, when the Argo buoys came into operation, the global oceans above 1900 m depth have, on average, warmed somewhat. The maximum warming (between the surface and 120 m depth) mainly affects oceans near the Equator, where the incoming solar radiation is at its maximum. Net cooling since 2004 is pronounced for the North Atlantic.

7. Data from tide gauges all over the world suggest that average global sea-level rise has been 1–1.5 mm/year, while the satellite-derived record suggests a rise of about 3.2 mm/year since 1992. There is still no widely accepted explanation for this marked difference.

8. Arctic and Antarctic sea ice extent since 1979 have exhibited opposite trends, decreasing and increasing, respectively. Superimposed on these overall trends, however, variations of shorter duration are also important to understand year-to-year variations. In the Arctic, a 5.3-year periodic variation is important, while for the Antarctic a variation of about 4.5 years duration is important. Both of these variations reached their minima simultaneously in 2016, which explains the simultaneous minimum in global sea ice extent. A shift towards larger ice extents in both hemispheres may have begun in 2017 (Figure 34), as predicted in 'State of the Climate 2016'.

9. The Northern Hemisphere snow cover extent has undergone important local and regional variations from year to year. The overall tendency since 1972, however, is towards overall stable snow extent conditions.
10. Tropical storm and hurricane accumulated cyclone energy (ACE) values since 1970 have displayed large variations from year to year, but no overall trend towards either lower or higher activity. The 2017 global tropical storm and hurricane accumulated cyclone energy (ACE) was above the level recorded for 2016 but was within the range experienced since 1970.

1 General overview

The focus in this report is on observations and not on output from models. All references and data sources are listed at the end of the report.

Air temperatures

The year 2017 was the first year after the strong 2015–2016 El Niño in the Pacific Ocean. Considering the entire temperature record since 1850/1880, it was a warm year, but cooler than 2016. In 2017 the average global temperature began dropping back towards the level characterising the years before the recent El Niño episode. The 2015–16 global surface temperature peak appears therefore to be primarily caused by this oceanographic phenomenon.

Many Arctic regions experienced record high temperatures in 2016, but in 2017 conditions were generally cooler. The Arctic temperature peak in 2016 may have been affected by ocean heat released from the Pacific Ocean during the 2015–16 El Niño and subsequently transported to the Arctic region.

Many of the diagrams in this report focus on the period from 1979 onwards, reflecting the commencement of the satellite era, with access to a wide range of observations with nearly global coverage, including temperature. These data now provide a detailed view of temperature changes over time at different altitudes in the atmosphere. Among other phenomena, these observations reveal that while the widely recognised lower troposphere temperature pause began around 2002, a similar stratospheric temperature plateau had already begun back in 1995; that is, several years before a similar temperature plateau started near the planet's surface. Somewhat surprisingly, little attention has so far been given to this aspect of global climate change.

An interesting difference prevails between average global air temperatures as estimated by surface stations (HadCRUT, NCDC and GISS) and by satellites (UAH and RSS). In the early part of the temperature record since 1979, the satellite-based temperatures were often – but not always – somewhat higher than the average temperature estimate derived from surface observations. Since 2004, however, the temperature estimate from surface stations has drifted away from the satellite-based estimate in a warm direction and is now on average about 0.1 °C higher, even though in 2017 one of the satellite records (RSS) was adjusted upwards.

Air temperatures measured near the planet's surface (surface air temperatures) are, however, still at the core of many climate deliberations, but the significance of any short-term warming or cooling recorded by surface air temperatures should not be overstated. Whenever Earth experiences a warm El Niño or a cold La Niña episode, major heat exchanges take place between the Pacific Ocean and the atmosphere above, eventually showing up as a signal in estimates of the global air temperature. However, this does not reflect similar changes in the total heat content of the atmosphere–ocean system. In fact, global net changes can be small, and such heat exchanges may chiefly reflect redistribution of energy between ocean and atmosphere. Evaluating the dynamics of ocean temperatures is therefore equally important as evaluating changes in surface air temperatures.

Oceans

The Argo program has now achieved 13 years of global coverage, growing from a relatively sparse global array of 1000 buoys in 2004 to more than 3000 in late 2007, and up to the present figure of about 3800 buoys. These Argo floats have provided a unique ocean temperature data set for depths down to 1900 m. Although the oceans are much deeper than 1900 m, and the Argo data series still is relatively short, several thought-provoking features are now emerging from these observations.

Since 2004 the upper 1900 m of the oceans have been warming on average. The maximum warming (0.23–0.08°C) affects the uppermost 100 m, mainly in regions near the Equator, where the greatest amount of solar radiation is received. At greater depths, a small net warming of about 0.02°C has occurred between 2004 and 2017, according to the Argo buoys.

This global average oceanic development since 2004 is well reflected in equatorial oceans – between 30°N and 30°S – which, due to the spherical shape of the planet, represent a huge surface area. Simultaneously, the northern oceans (55–65°N) have on average experienced a marked cooling down to 1400 m depth, and some warming at greater depths. The southern oceans (55–65°S) on average have seen slight warming at all depths since 2004. However, averages may be misleading, and quite often a better insight is gained by studying the details, which are considered later in this report.

Sea ice

In 2017, global sea ice cover extension remained well below the average for the satellite era (since 1979), but with a rising trend indicated. At the end of 2016 the global sea ice extent reached a marked minimum, at least partly caused by the operation of two different natural variation patterns that characterise sea ice in the Northern and Southern Hemispheres, respectively. Both variations had simultaneous minima in 2016, with consequences for the global sea ice extent. A trend in the opposite direction, towards higher ice extent at both poles, appears to have begun during 2017, as was predicted in 'State of the Climate 2016'.

Snow cover

Variations in the global snow cover extent were mainly caused by changes in the Northern Hemisphere, where all the major land areas are located. The Southern Hemisphere snow cover extent is essentially controlled by the Antarctic ice sheet, and is therefore relatively stable. The Northern Hemisphere average snow cover extent has also been stable since the onset of satellite observations, although local and regional interannual variations may be large. Considering seasonal changes since 1979, the Northern Hemisphere snow cover autumn extent is slightly increasing, the mid-winter extent is approximately stable, and the spring extent is slightly decreasing. In 2017, the Northern Hemisphere snow cover extent was somewhat higher than in 2016, especially during winter, spring and summer.

Sea level

Global sea levels are monitored by satellite altimetry and by direct measurements from tide-gauges along coasts. While the satellite-derived record suggests a global sea level rise of about 3.2 mm per year, data from tide gauges all over the world suggest a stable, average global sea-level rise of less than 1.5 mm per year. There is still no widely accepted explanation for this marked difference. What remains important, however, is that for local coastal planning the data from traditional tide gauges are more important, as is detailed later in this publication.

Tropical storms and hurricanes

The 2017 global tropical storm and hurricane accumulated cyclone energy (ACE) was above the level recorded for 2016 but was within the range experienced since 1970. The ACE data series display a variable pattern over time, but without any clear trend towards higher or lower values. A longer ACE series for the Atlantic basin (since 1850), however, suggests a natural cycle of about 60 years duration for global tropical storm and hurricane ACE.

2 The spatial pattern of global surface air temperatures

On average, the global surface air temperature for year 2017 was near the average of the past ten years (Figure 1). The previous two years, 2015 and 2016, were affected by the recent El Niño episode playing out in the Pacific Ocean and culminating in early 2016. In 2017 the global surface air temperature was slowly dropping back towards the pre-2015–16 level. The Northern

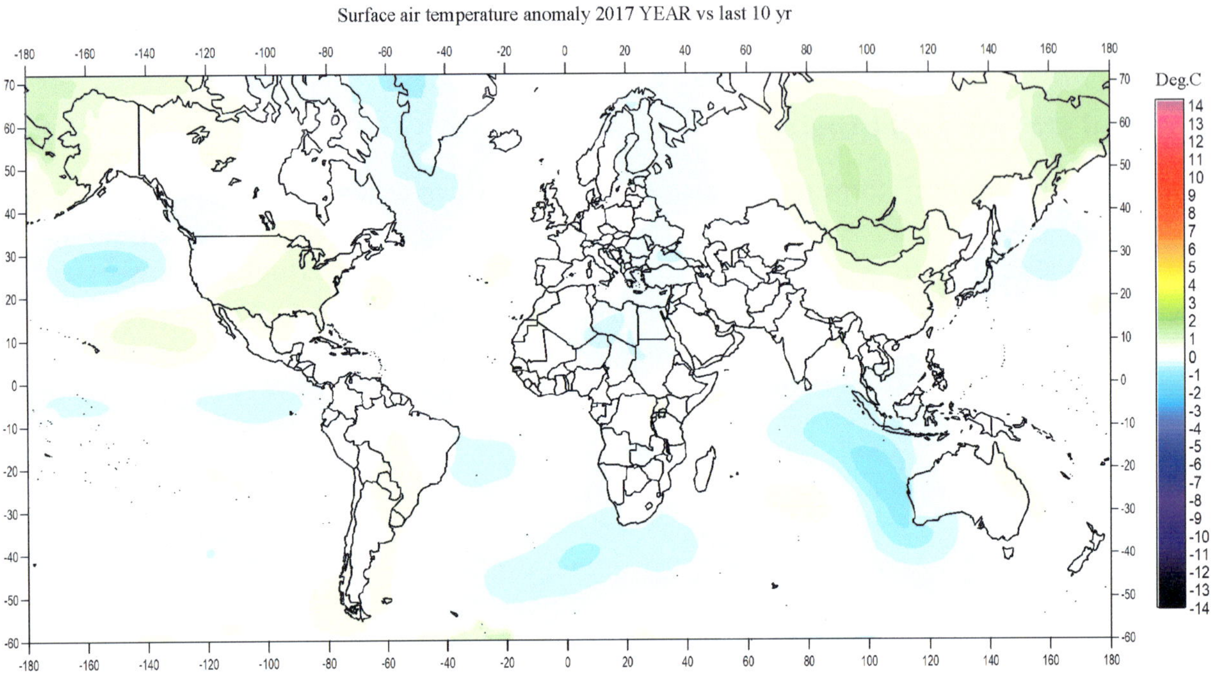

Figure 1: Surface air temperatures compared to the average for the previous 10 years. Green-yellow-red colours indicate areas with higher temperature than the average, while blue colours indicate lower than average temperatures. Data source: Goddard Institute for Space Studies (GISS) using $ERSST_{v4}$ ocean surface temperatures.

Hemisphere was characterised by regional temperature contrasts but was generally relatively warm in the Alaska-Siberia regions. Also, most of North America and China had temperatures somewhat above average for the previous 10 years. In contrast, the North Atlantic region with Greenland and Europe was generally relatively cold, as was the northern Pacific.

Near the Equator, temperature conditions were generally near the average for the previous 10 years. In the eastern Pacific, however, temperatures were relatively cold, conceivably reflecting the onset of La Niña.

In the Southern Hemisphere, surface air temperatures were near or below the average for the previous 10 years. In particular, the Indian Ocean west of Australia and most of the South Atlantic had temperatures somewhat below the average.

In the central Arctic, in contrast to 2016, most regions in 2017 had below average temperatures (Figure 2a). This development was especially pronounced within the Canada-Greenland-

Europe Arctic sectors.

The Antarctic continent was mainly characterised by above-average temperatures in 2017, with only part of East Antarctica having temperatures below the average for the past 10 years (Figure 2b).

In summary, in 2017, global average air temperatures are approaching the level characterising the years leading up to the recent 2015–16 El Niño episode. Thus, the global surface air temperature peak of 2015–16 appears mainly to be caused by this Pacific oceanographic phenomenon.

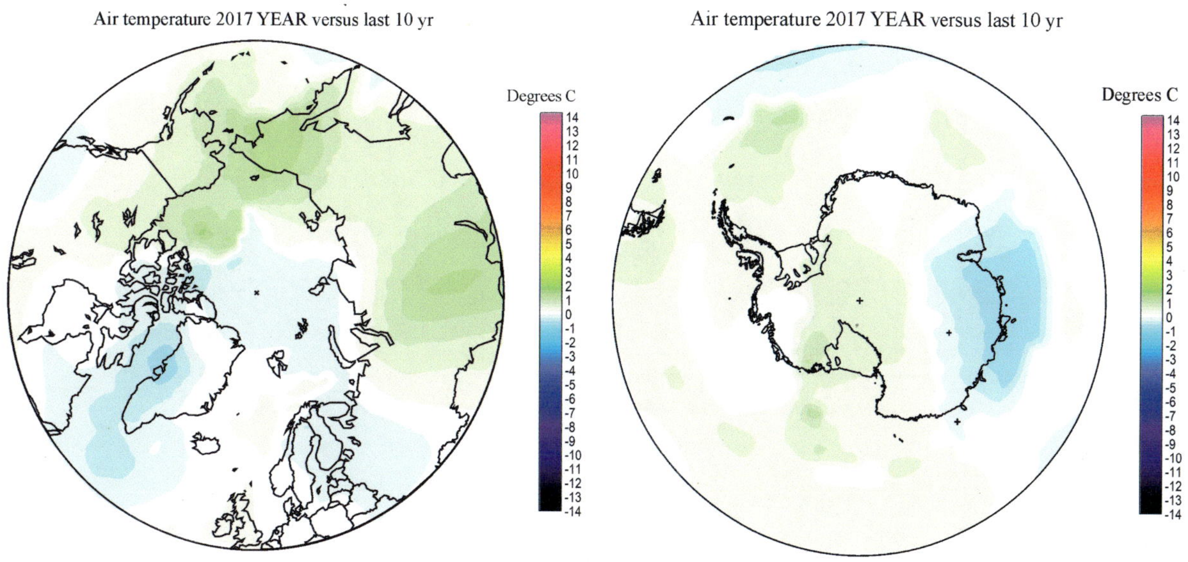

(a) (b)

Figure 2: Polar surface air temperatures compared to the average for the previous 10 years. Green-yellow-red colours indicate areas with higher temperature than the average, while blue colours indicate lower than average temperatures. Data source: Goddard Institute for Space Studies (GISS) using $ERSST_{v4}$ ocean surface temperatures.

3 Global monthly lower troposphere air temperature since 1979

Both satellite records for the lower troposphere temperature clearly show the temperature spike associated with the 2015–16 El Niño, and the subsequent drop towards the level characterising the period before the recent El Niño episode (Figure 3). Comparison of the latest (December 2017) record and the May 2015 record (red) shows that only a few small adjustments have been made to the University of Alabama, Huntsville (UAH) series, while the Remote Sensing Systems (RSS) series was recently subject to a large adjustment towards higher temperatures from 2002 and onwards; about +0.1 °C. This adjustment was introduced in 2017.

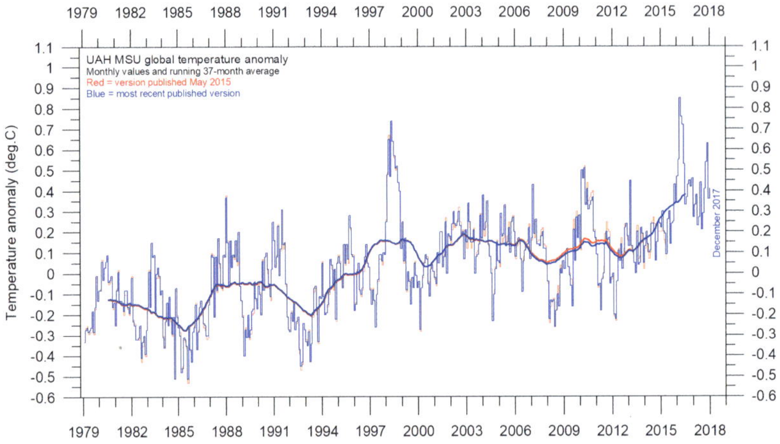

(a)

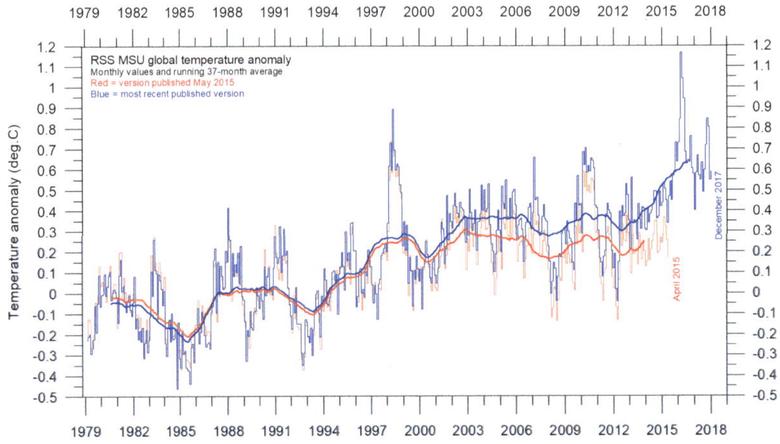

(b)

Figure 3: Global monthly average lower troposphere temperatures since 1979, representing conditions at about 2 km altitude: (a) UAH, (b) RSS.

The thick lines are the simple running 37-month averages, approximately corresponding to a running three -year average.

Temporal diagrams of global monthly lower troposphere temperatures since 1979 clearly shown the effects of the El Niños of 1998, 2010 and 2016 and also the tendency for many El Niños to culminate during the Northern Hemisphere winter (Figure 4). In these temporal diagrams the annual time axis is shown vertically, and the monthly time axis horizontally.

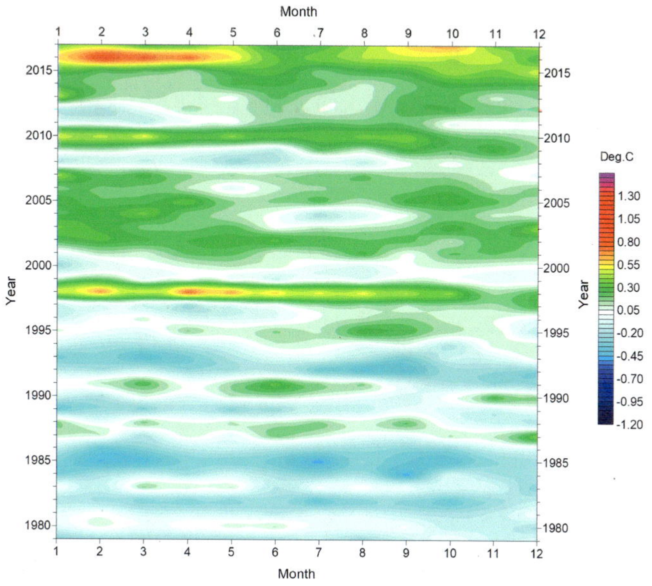

(a)

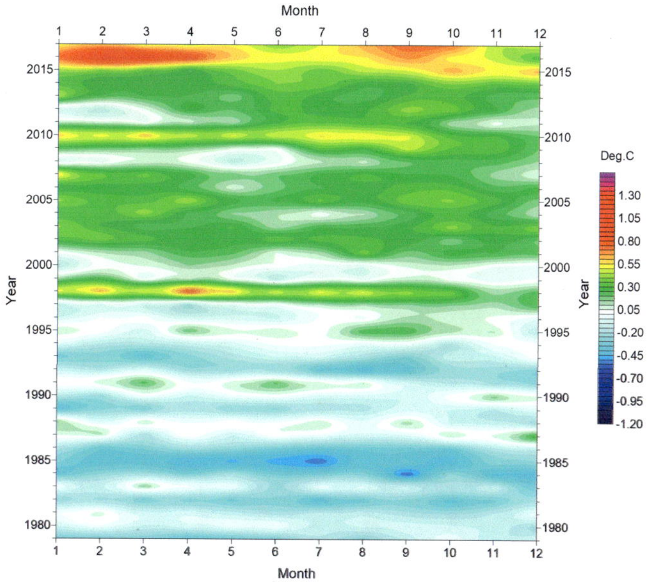

(b)

Figure 4: Temporal diagrams of global monthly lower troposphere temperatures since 1979.
(a) UAH record; (b) RSS record. As the different temperature databases are using different reference periods, the series have been made comparable by setting their individual 30-year averages (1979–2008) at zero.

4 Global mean annual lower troposphere air temperature since 1979

Annual global mean temperatures for the lower troposphere are shown in Figure 5. The 30-year average (1979–2008) is used as zero value in these diagrams, to enable easy comparison. The year 2017 is cooler than 2016.

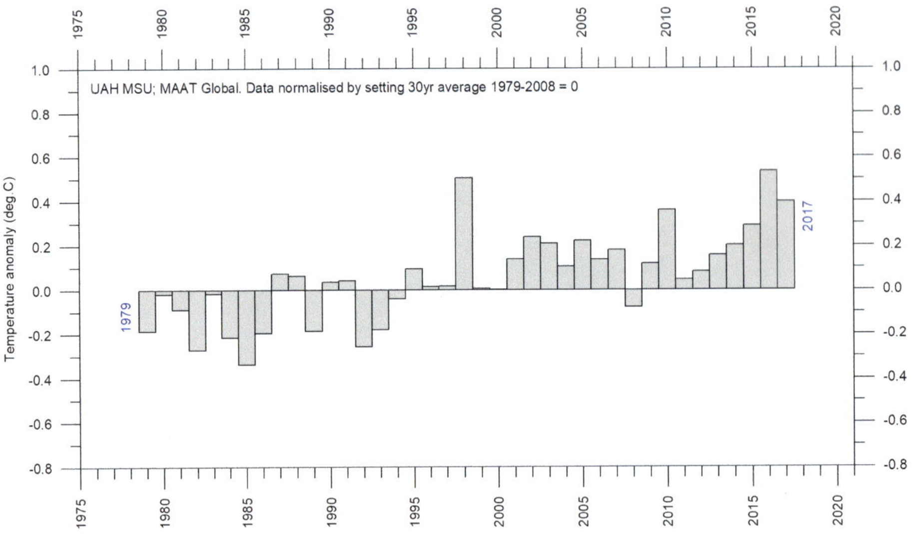

(a)

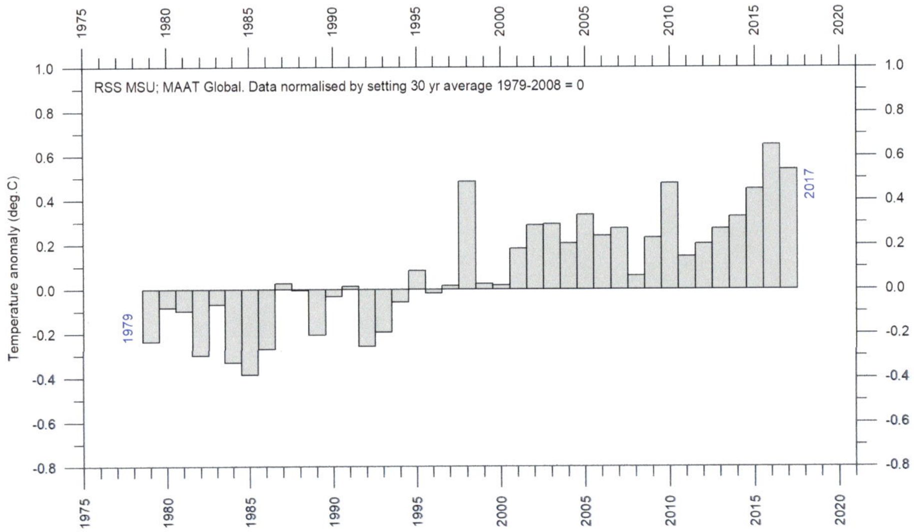

(b)

Figure 5: Global mean annual lower troposphere air temperatures since 1979.
(a) UAH record (b) RSS record.

5 Global monthly surface air temperatures since 1979

All three surface air temperature records clearly show the temperature spike associated with the 2015–16 El Niño (Figure 6). At the end of 2017, however, the global temperature is again moving towards the general level characterising the period before the recent El Niño episode, according to all three records.

The comparison between the most recent (December 2017) record and the May 2015 record (red) shows that few adjustments has since been introduced in the HadCRUT record, while numerous and relatively large changes have been introduced in the NCDC and GISS records. All three surface records however confirm that the recent El Niño episode culminated in early 2016.

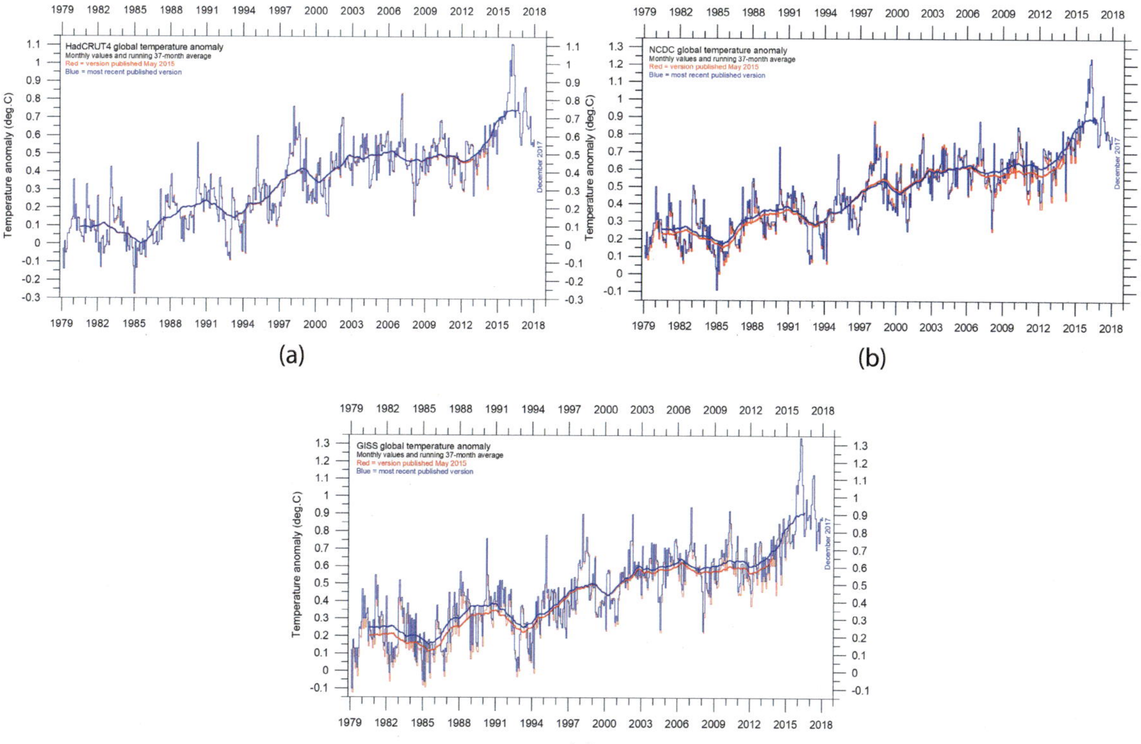

Figure 6: Global monthly average surface air temperature since 1979.
(a) According to HadCRUT, a cooperative effort between the Hadley Centre for Climate Prediction and Research and the University of East Anglia's Climatic Research Unit (CRU), UK. (b) According to the National Climatic Data Center (NCDC), USA; (c)according to the Goddard Institute for Space Studies (GISS) USA. The thick line is the simple running 37-month average, nearly corresponding to a running 3-year average.

The temporal evolution of these temperatures is shown in Figure 7, with the annual time axis shown vertically, and the monthly time axis horizontally. The culmination of the recent El Niño episode in early 2016 is clearly seen in all three records.

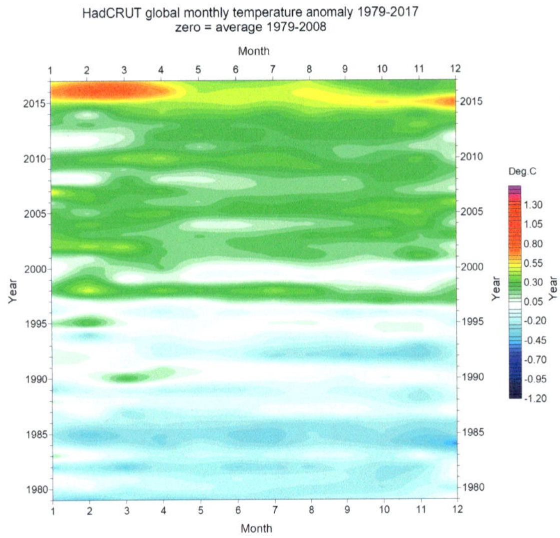

(a)

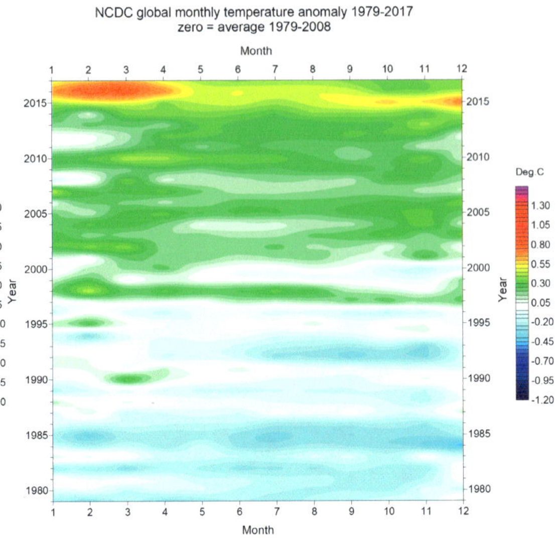

(b)

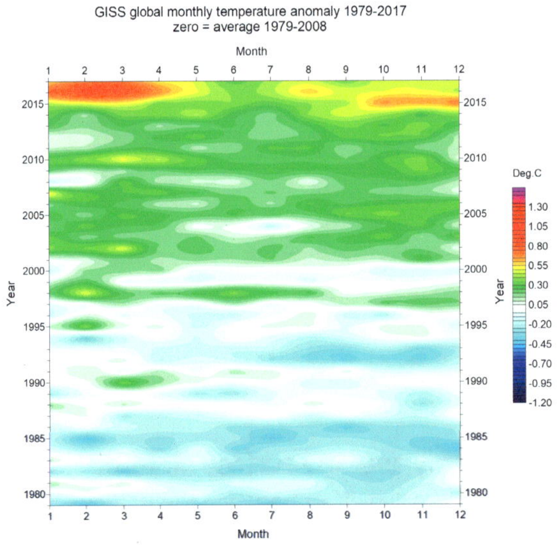

(c)

Figure 7: Temporal diagrams showing global monthly lower troposphere temperatures since 1979.

(a) HadCRUT, (b) NCDC and (c) GISS. As the different temperature databases are using different reference periods, the series have been made comparable by setting their individual 30-year average 1979–2008 as zero value.

6 Global mean annual surface air temperature

Annual global mean surface temperature records are shown in Figure 8. All three average surface air temperature estimates show the year 2016 to be the warmest on record. The year 2016 was influenced by the recent strong El Niño episode.

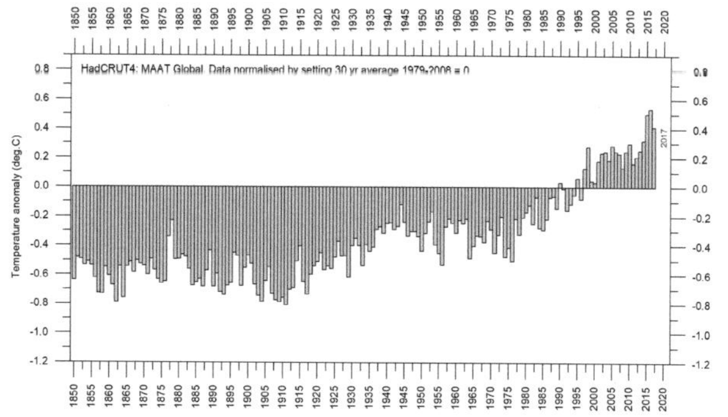

(a)

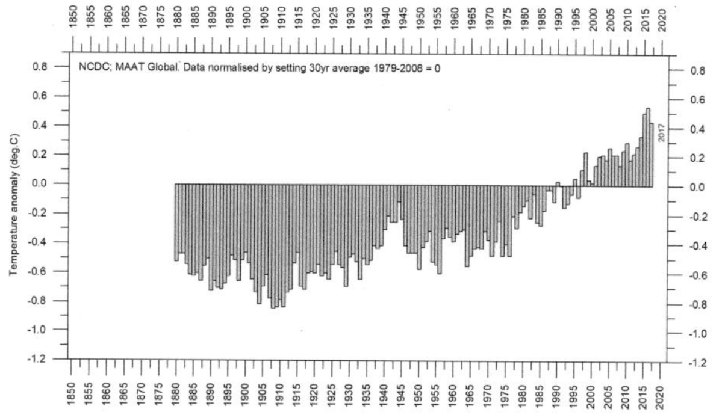

(b)

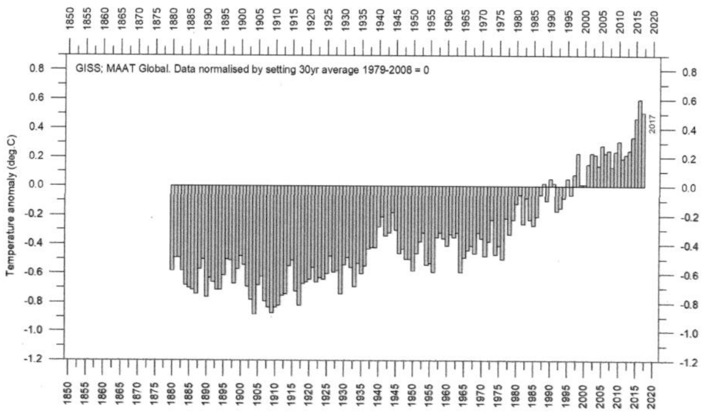

(c)

Figure 8: Global mean annual surface air temperatures since 1850.
According to (a) Hadley CRU, (b) NCDC, (c) GISS.

7 Comparing surface air temperatures with temperatures recorded by satellites

There remains an interesting difference between global air temperatures estimated by surface stations and satellites, respectively, as illustrated by Figures 9 and 10. In the early part of the record since 1979, the satellite-based temperatures were often somewhat higher than global estimates derived from surface observations. Since 2004, however, the temperature estimate from surface stations has drifted away from the satellite-based estimate in a warm direction. The 2017 adjustment of the RSS satellite record has reduced this difference compared to the situation in previous years.

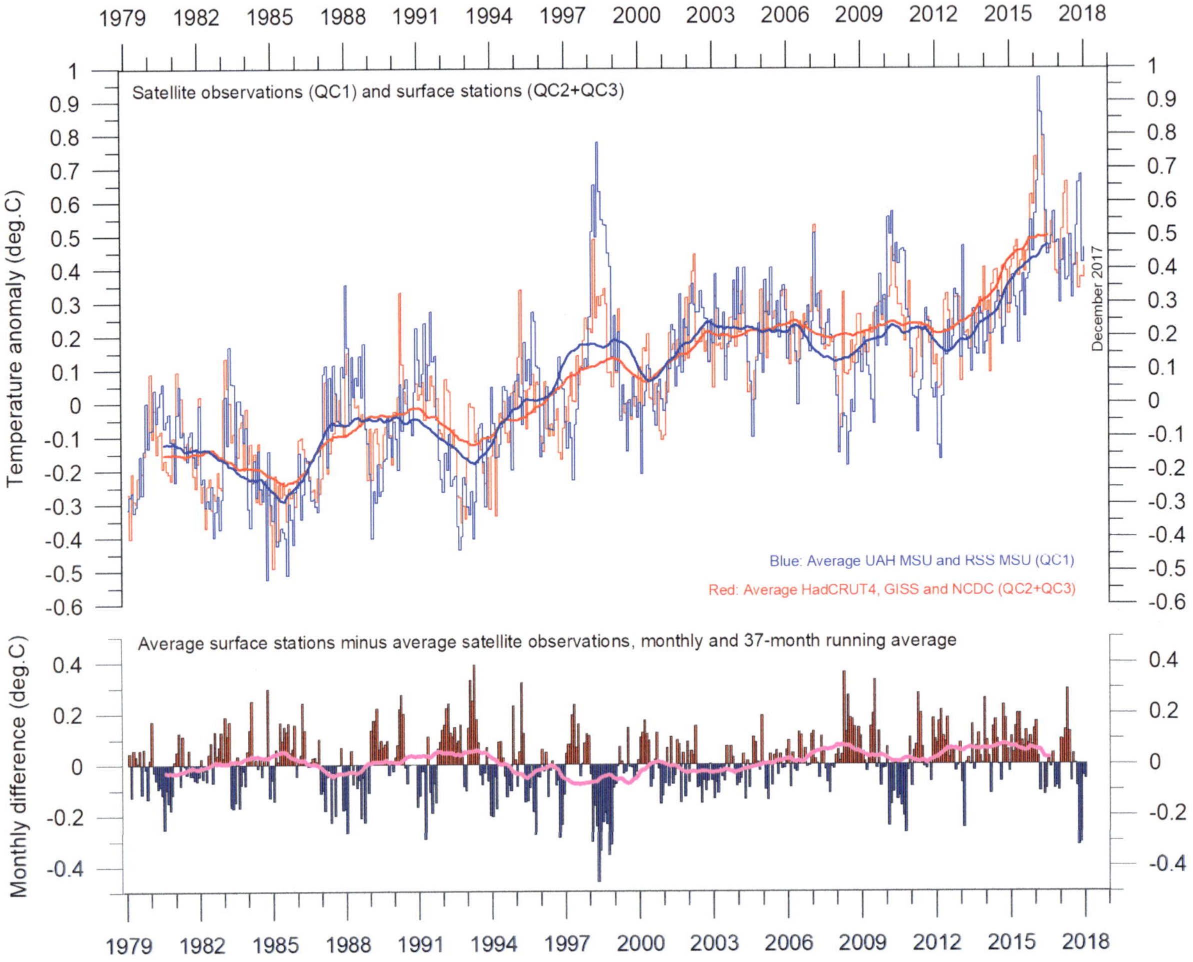

Figure 9: Plot showing the average of monthly global surface air temperature estimates (HadCRUT, NCDC and GISS) and satellite-based temperature estimates (UAH and RSS).

The thin lines indicate the monthly value, while the thick lines represent the simple running 37-month average, nearly corresponding to a running 3-year average. The lower panel shows the monthly difference between surface air temperature and satellite temperatures. As the base period differs for the different temperature estimates, they have all been normalised by comparing to the average value of 30 years from January 1979 to December 2008.

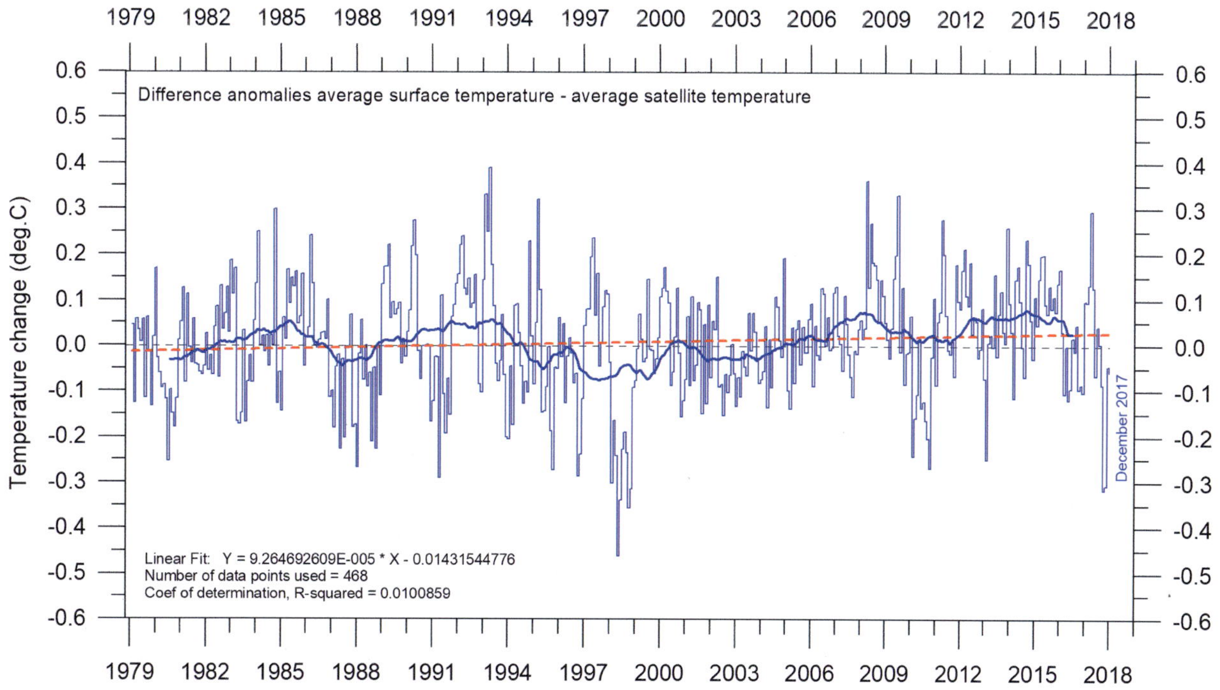

Figure 10: Global monthly average surface air temperature (HadCRUT, NCDC, GISS) minus global monthly average lower troposphere temperature (UAH, RSS) since 1979.

The thin blue line shows the monthly temperature difference between anomalies calculated for surface and lower troposphere observations, respectively. The thick blue line is the simple running 37-month average, nearly corresponding to a running 3-year average. The dotted red line is the linear fit line, statistics of which is presented in the lower left corner of the diagram. As the base period differs for the different temperature estimates, they have all been normalised by comparing to the average value of 30 years from January 1979 to December 2008.

8 Comparing temperature change over land and oceans; lower troposphere air temperature

Since 1979 lower troposphere temperatures have increased more over land, compared to over oceans (Figure 11). Especially since about 2006 temperatures recorded over land has been consistently higher than above oceans. There may be several reasons for this, such as, e.g., variations in insolation, cloud cover and land use, affecting mainly land areas.

This development has apparently taken place roughly in concert with the above-mentioned drift of surface observations towards higher temperatures, compared to satellite-based temperatures (see previous paragraph).

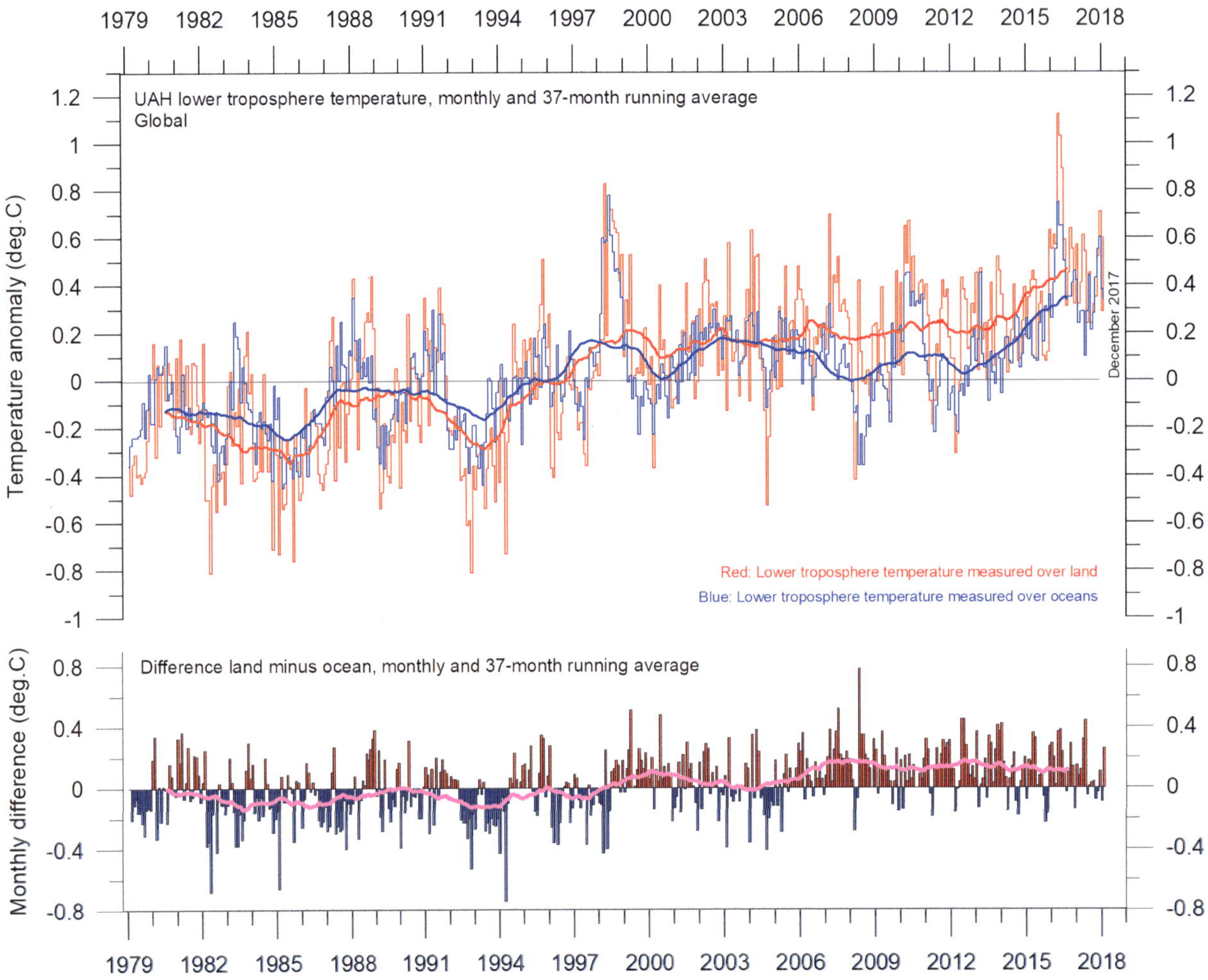

Figure 11: Global monthly average lower troposphere temperature over land and ocean since 1979.

Red, land; blue, oceans. Data according to UAH. The thin lines represent the monthly average, and the thick line the simple running 37-month average, nearly corresponding to a running 3-year average.

9 Comparing atmospheric temperatures from surface to 17 km altitude

The temperature variations recorded in the lowermost troposphere are generally reflected at higher altitudes, up to at least 5–6 km altitude. The overall temperature plateau since about year 2002 is found at all these altitudes (Figure 12). At higher altitudes (about 10 km), near the tropopause, the pattern of variations recorded lower in the atmosphere can still be recognised, but for the duration of the record (since 1979) there has been no trend towards higher or lower temperatures.

Higher in the atmosphere, in the stratosphere, at 17 km altitude, two pronounced temperature spikes are visible before the turn of the century. Both spikes can be related to major volcanic eruptions, as indicated in the diagram. Ignoring these spikes, until about 1995 the stratospheric temperature record shows a persistent decline, ascribed by some scientists to the effect of more and more heat being trapped by CO_2 in the troposphere below. However, this temperature decline ends around 1995–96, and a long temperature plateau has since then characterised the stratosphere. Thus, the stratospheric temperature pause initiated 5–7 years before a similar pause commenced in the lower troposphere.

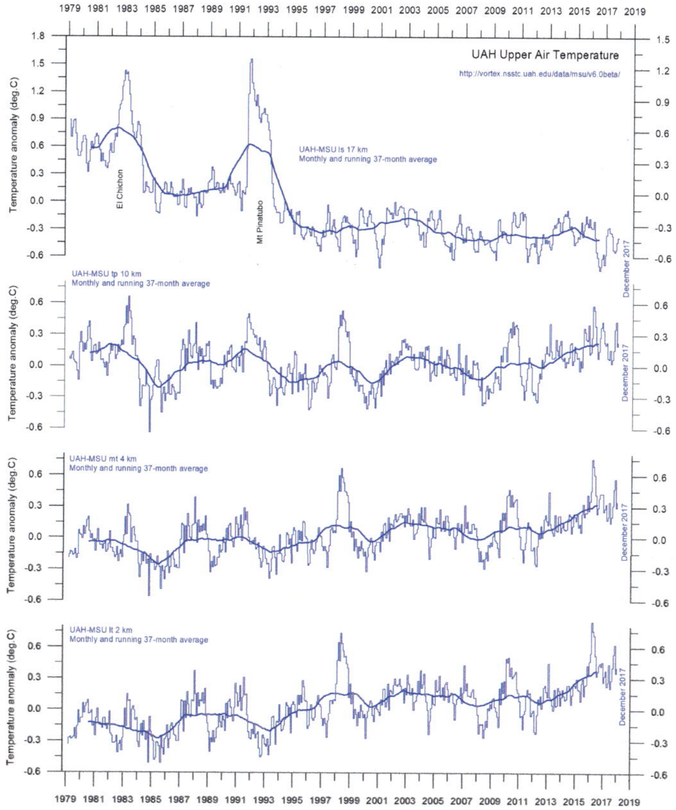

Figure 12: Global monthly average temperature in different altitudes.
Data according to UAH. The thin lines represent the monthly average, and the thick line the simple running 37-month average, nearly corresponding to a running 3-year average.

10 Atmospheric greenhouse gases; water vapour and carbon dioxide

Water vapour

Water vapour is the most important greenhouse gas in the troposphere. The highest concentration is found within a latitudinal range from 50°N to 60°S. The two polar regions of the troposphere are comparatively dry.

The specific atmospheric humidity is seen to be stable or slightly increasing up to about 4–5 km altitude (Figure 13). At higher levels in the troposphere (about 9 km), the specific humidity has been decreasing for the duration of the record (since 1948), but with shorter variations superimposed on the falling trend. A Fourier frequency analysis (not shown here) shows these variations to be influenced especially by a periodic variation of about 3.7 years' duration.

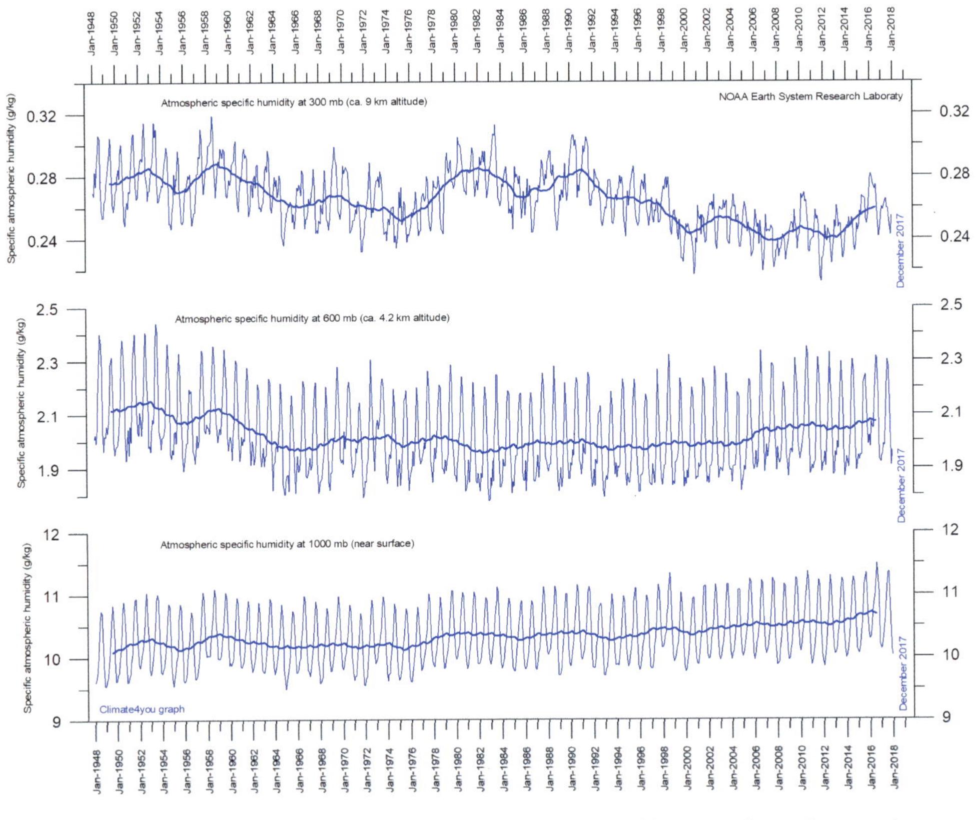

Figure 13: Specific atmospheric humidity (g/kg) at three different altitudes in the troposphere since January 1948.

The thin blue lines show monthly values, while the thick blue lines show the running 37-month average (about 3 years). Data source: Earth System Research Laboratory (NOAA).

The persistent decrease in specific humidity at about 9 km altitude is interesting, as this altitude roughly corresponds to the level where the theoretical temperature effect of atmospheric CO_2 is expected initially to play out.

Carbon dioxide

Carbon dioxide (CO_2) is an important greenhouse gas, although less important than is water vapour. For the duration of the Mauna Loa record (since 1958), an increasing trend is clearly visible, with an annual cycle superimposed (Figure 14). At the end of 2017, the amount of atmospheric CO_2 is slightly above 400 ppm. Usually, CO_2 is considered as a relatively well-mixed gas in the troposphere. The 12-month change in tropospheric CO_2 has been increasing from

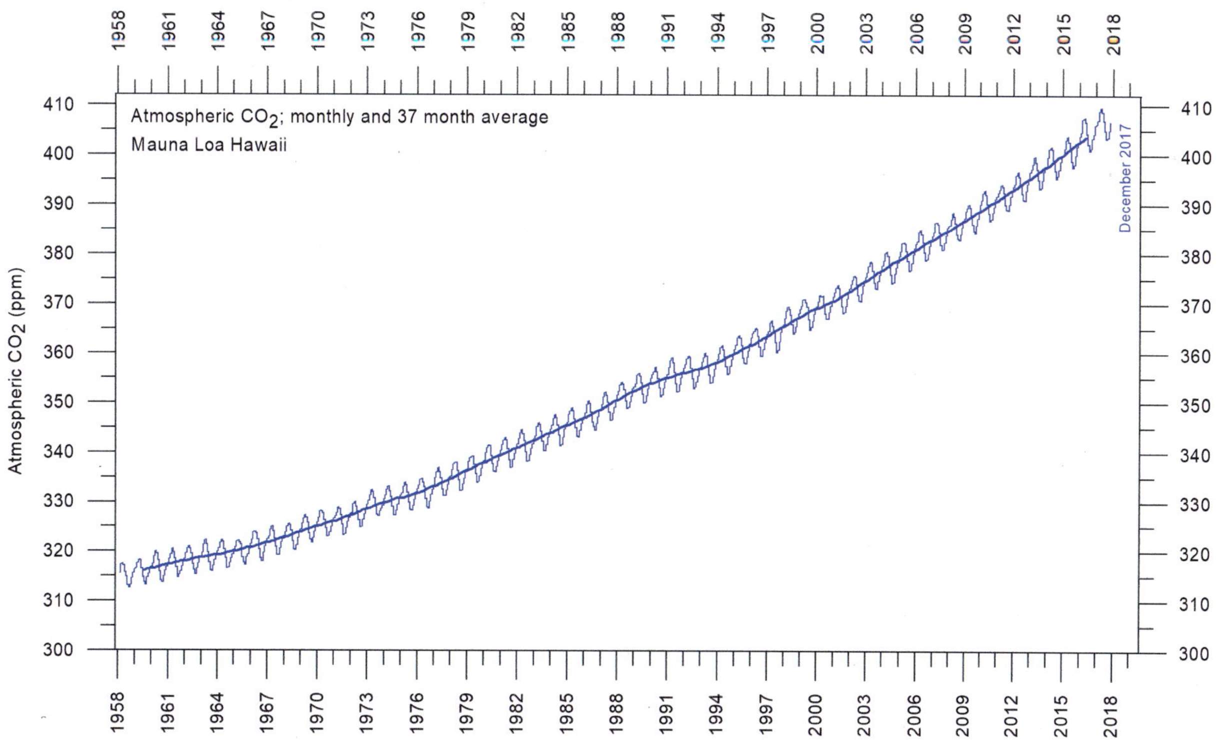

Figure 14: Monthly amount of atmospheric CO_2 since March 1958.
As measured at the Mauna Loa Observatory, Hawaii. The thin line shows the monthly values, while the thick line is the simple running 37-month average, nearly corresponding to a running 3-year average.

about +1 ppm/year in the early part of the record, to more than +3 ppm/year towards the end of the record (Figure 15). A Fourier frequency analysis (not shown here) shows the 12-month change of tropospheric CO_2 to be influenced especially by periodic variations of 2.5 and 3.8 years' duration, respectively.

It is instructive to consider the variation of the annual change rate of atmospheric CO_2 together with the annual change rates for the global air temperature and global sea surface temperature (Figure 16). All three change rates clearly vary in concert, but with sea surface temperatures leading a few months ahead of the global temperature and change rates for atmospheric CO_2 lagging 11–12 months behind the sea surface temperature change rates.

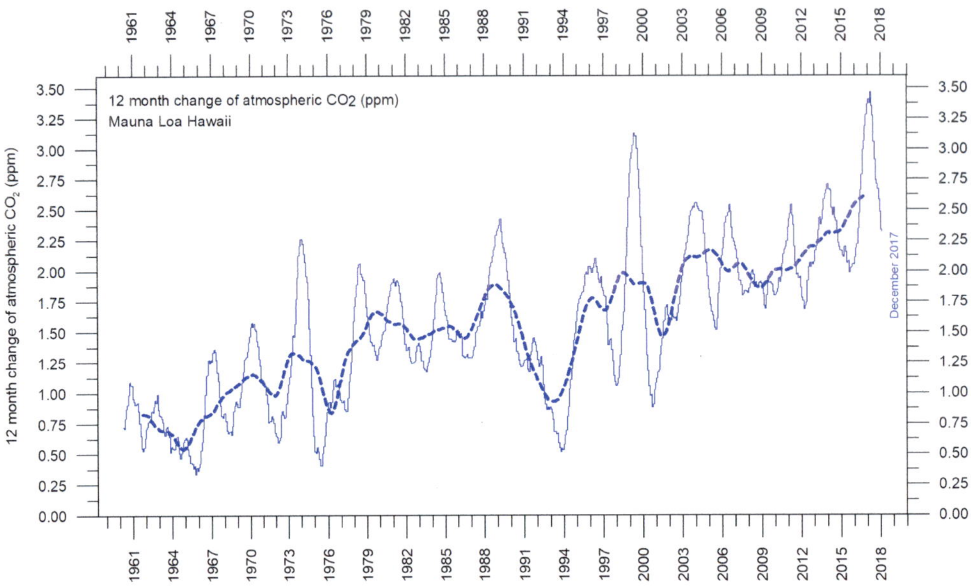

Figure 15: Annual (12 month) growth rate (ppm) of atmospheric CO_2 since 1959.

Calculated as the average amount of atmospheric CO_2 during the last 12 months, minus the average for the preceding 12 months. The graph is based on data measured at the Mauna Loa Observatory, Hawaii. The thin blue line shows the value calculated month by month, while the dotted blue line represents the simple running 3-year average.

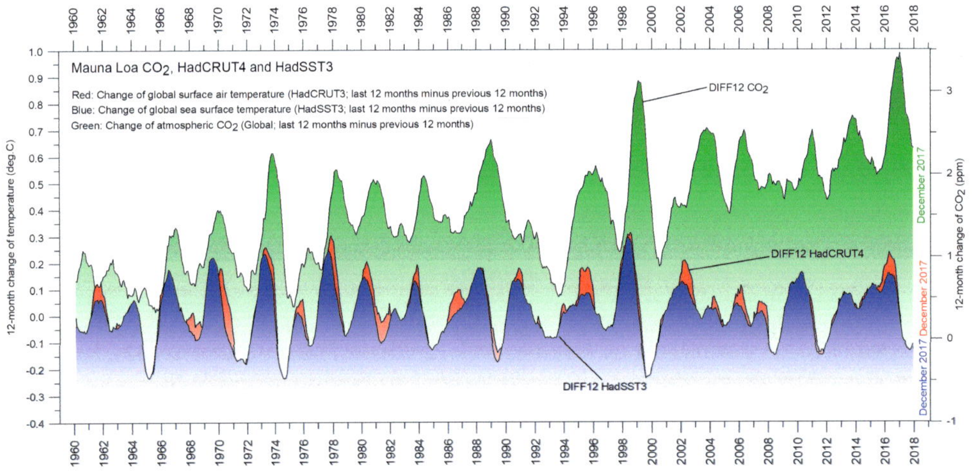

Figure 16: 12-month change of global atmospheric CO_2 concentration (Mauna Loa; green), global sea surface temperature (HadSST3; blue) and global surface air temperature (HadCRUT4; red dotted).

All graphs are showing monthly values of DIFF12, the difference between the average of the last 12 months and the average for the previous 12 months for each data series.

11 Zonal surface air temperatures

Figure 17 shows that the 'global' warming experienced after1980 dominantly was a Northern Hemisphere phenomenon, and mainly took place as a marked change between 1994 and 1999. This apparently rapid temperature change is, however, influenced by the Mount Pinatubo eruption 1992–93 and the following 1997 El Niño.

Figure 17 further shows the temperature effects of the equatorial El Niños in 1997 and 2015–16 spreading to higher latitudes in both hemispheres. The 2015–16 El Niño temperature effect was, however, mainly recorded in the Northern Hemisphere, and only to lesser degree in the Southern Hemisphere.

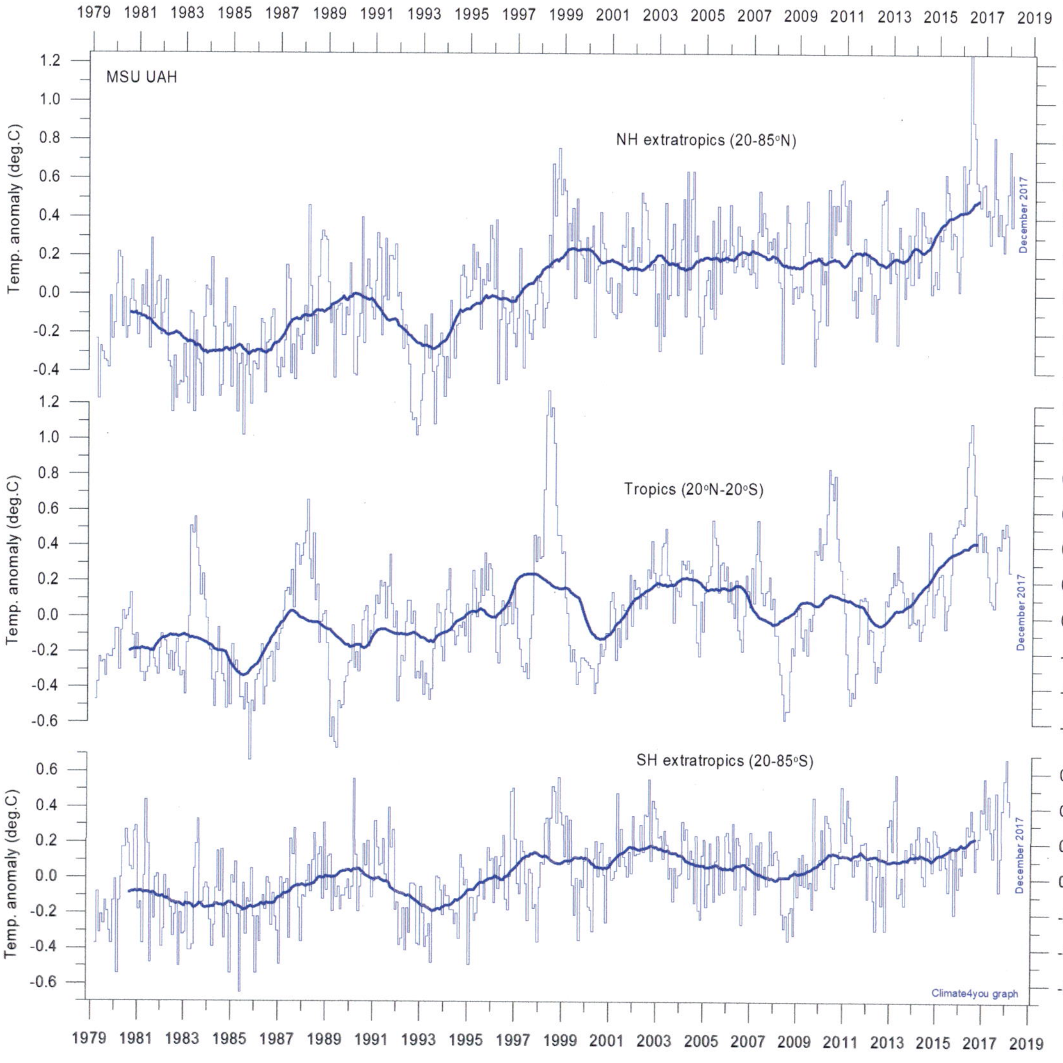

Figure 17: Global monthly average lower troposphere temperature since 1979 for the tropics and the northern and southern extratropics.

Data according to UAH. Thick lines are the simple running 37-month average, nearly corresponding to a running 3-year average.

12 Polar air temperatures

In the Arctic region, warming commenced around 1995, but faded somewhat out since about 2005 (Figure 18). In 2016, however, temperatures peaked for several months, presumably because of oceanic heat given off to the atmosphere during the recent El Niño 2015–16 (see also previous diagram) and subsequently advected to higher latitudes.

In the Antarctic region, temperatures have remained almost stable since the onset of the satellite record (in 1979). In 2016 a small temperature peak visible in the monthly record may be interpreted as the subdued effect of the recent El Niño episode.

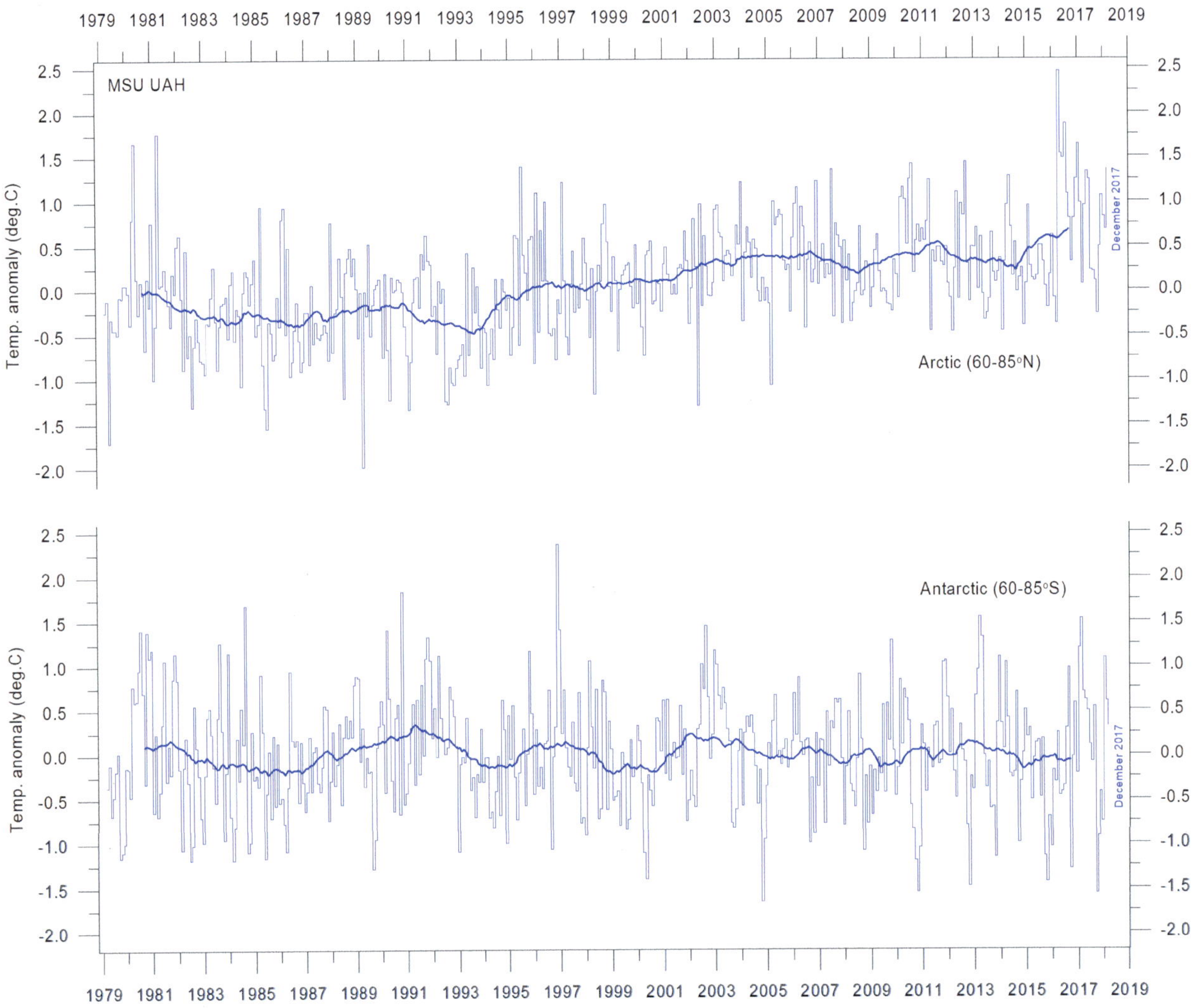

Figure 18: Global monthly average lower troposphere temperature since 1979 for the North Pole and South Pole regions.

Data according to UAH. Thick lines are the simple running 37-month average, nearly corresponding to a running three-year average.

13 Sea surface temperature anomaly at the end of the years 2015, 2016 and 2017

The three maps in Figure 19 show the recent El Niño in December 2015, compared to the post El Niño situation about one year later (December 2016), and the apparently coming La Niña (December 2017). In this way the three diagrams nicely illustrate the transformation from a full El Niño (end of 2015) to a beginning La Niña (end of 2017).

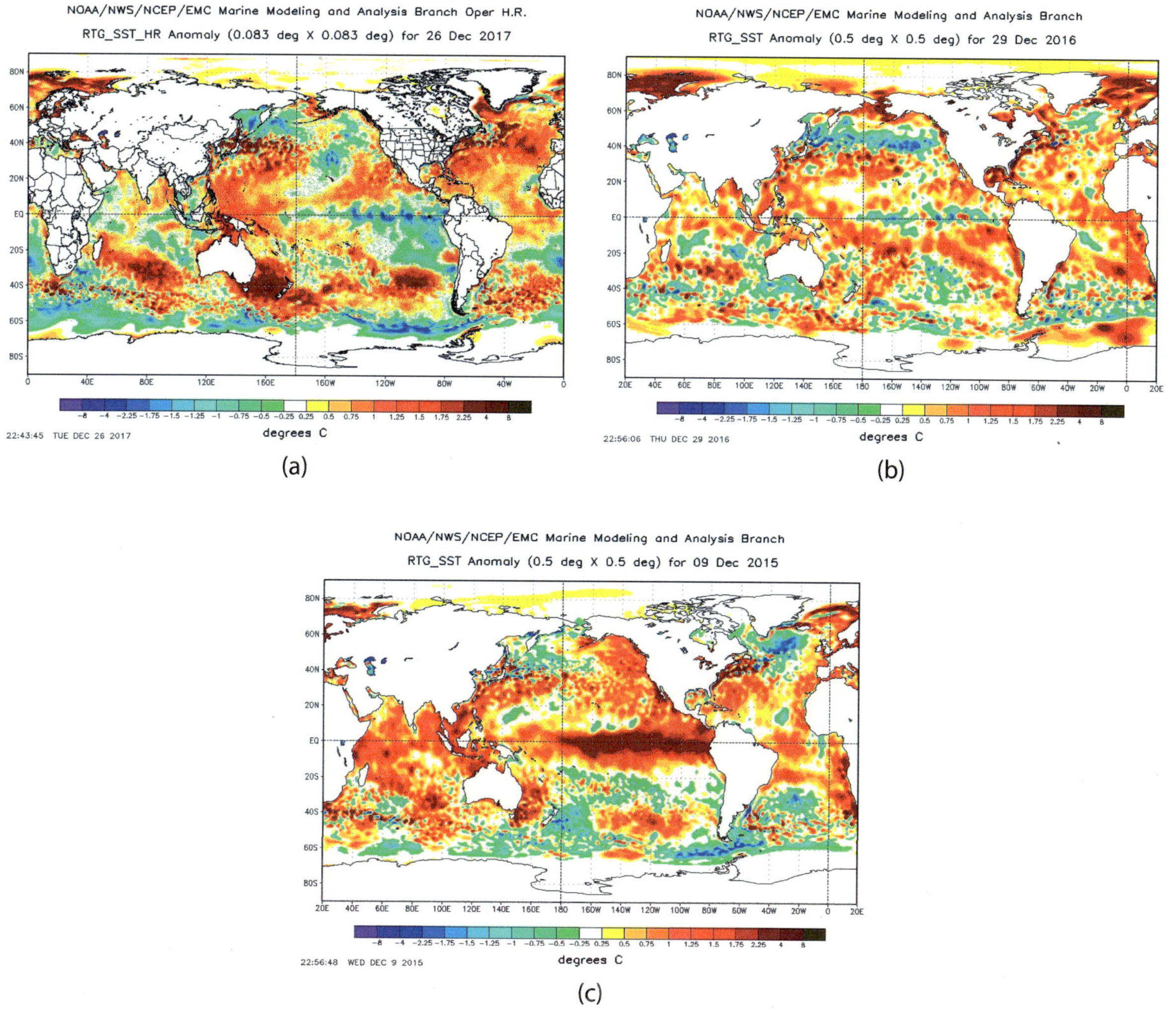

Figure 19: Sea surface temperature anomalies for December: (a) 2017, (b) 2016 and (c) 2015. (°C) The maps show the current anomaly (deviation from normal) of the surface temperature of Earth's oceans. White represents land areas. White shows the extent of sea ice. Map source: National Centers for Environmental Prediction (NCEP).

This is also illustrated by Figure 20. The recent 2015–16 El Niño episode is among the strongest since the beginning of the record in 1950. Considering the entire record, however, recent variations between El Niño and La Niña episodes do not appear abnormal in any way.

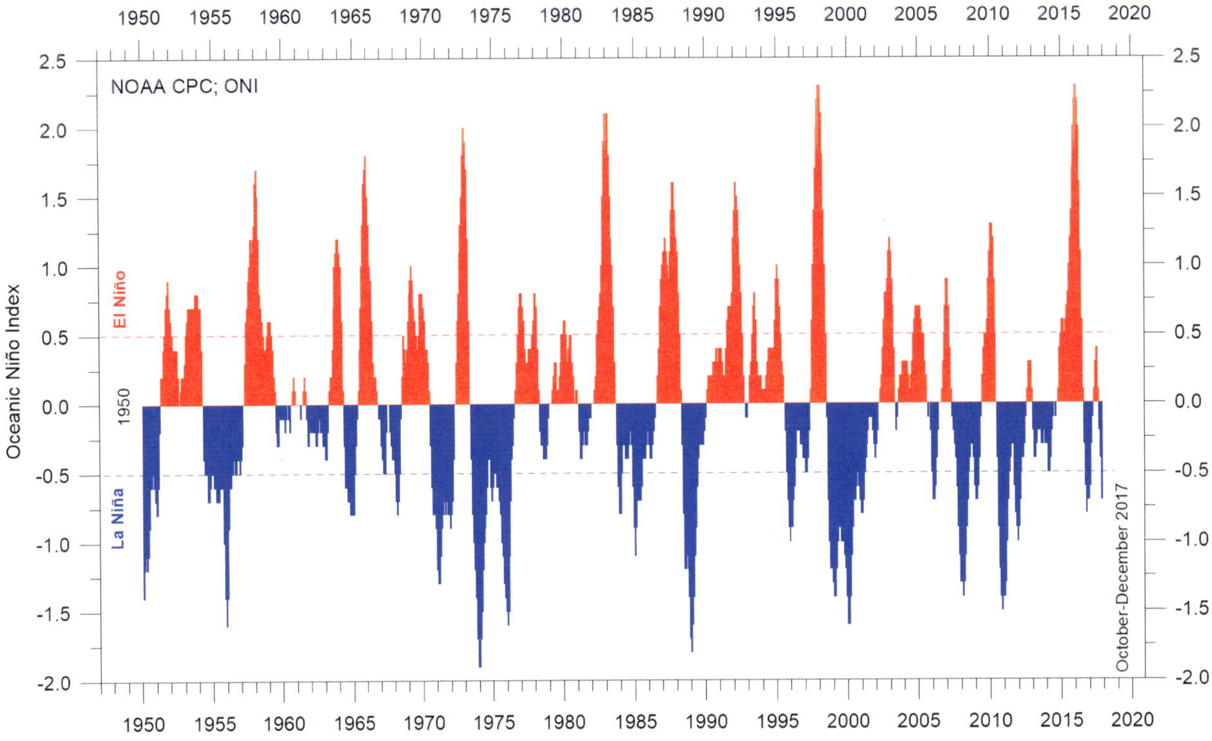

Figure 20: Warm and cold episodes for the Oceanic Niño Index (ONI).

ONI is defined as three-,month running mean of ERSSTv4 SST anomalies in the Niño 3.4 region (5°N-5°S, 120°170°W). Anomalies are centred on 30-year base periods updated every 5 years.

14 Global ocean average temperatures to 1900 m depth

Figure 21 is based on observations by Argo floats (Roemmich and Gilson 2009) and is a summary of ocean temperatures in different latitude bands. It shows that, on average, the temperature of the global oceans down to 1900 m depth has been increasing since about 2011, but not previously. Furthermore, it is seen that this increase is predominantly due to oceanic changes occurring near the Equator, between 30°N and 30°S. Near the Arctic, north of 55°N, ocean temperatures are decreasing. Near the Antarctic, south of 55°S, temperatures are essentially stable. At most latitudes, a clear annual rhythm is seen.

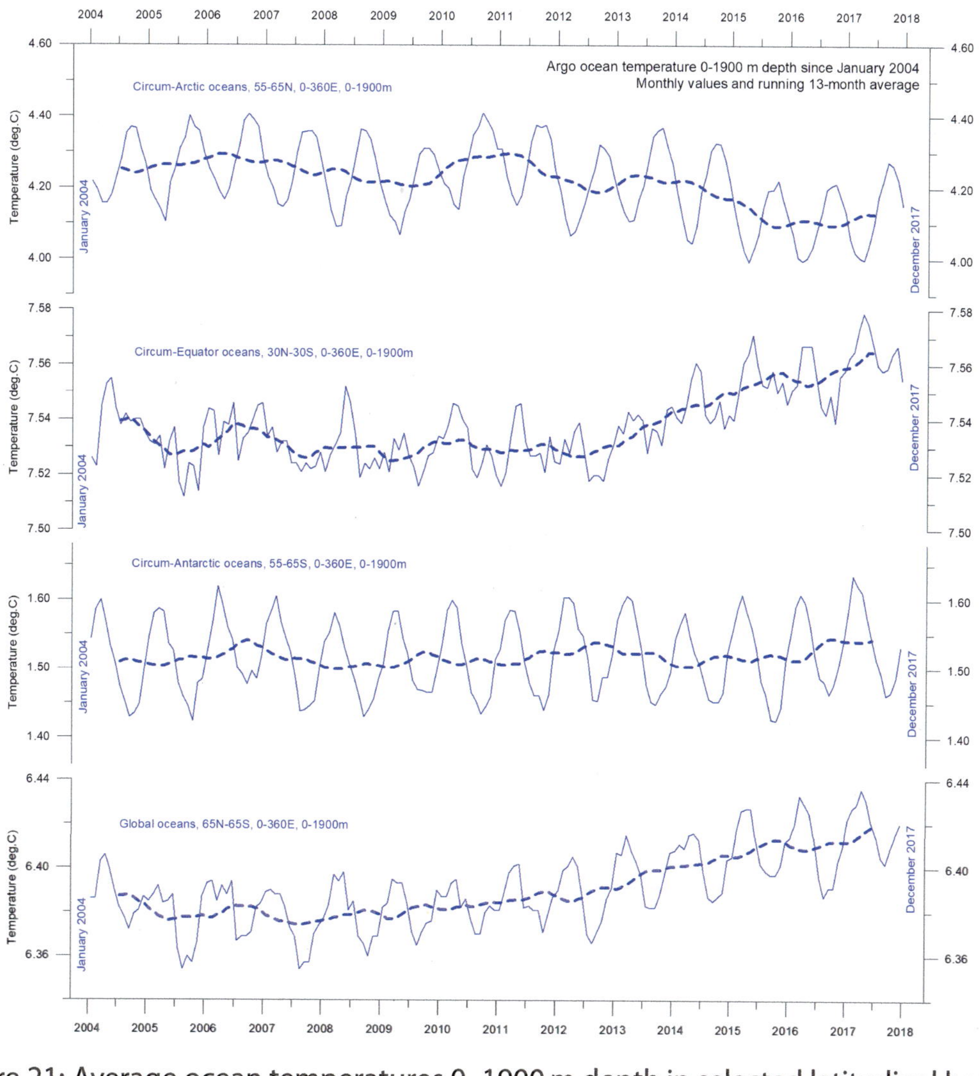

Figure 21: Average ocean temperatures 0–1900 m depth in selected latitudinal bands. Argo data. The thin line shows monthly values and the thick line shows the running 13-month average. Source: Global Marine Argo Atlas.

15 Global ocean temperatures at different depths

Figure 22 above shows global average oceanic temperatures at different depths. The annual rhythm can be traced down to about 100 m depth. In the uppermost 100 m, temperatures have increased since about 2011, while at 200–400 m depth temperatures have experienced little change only.

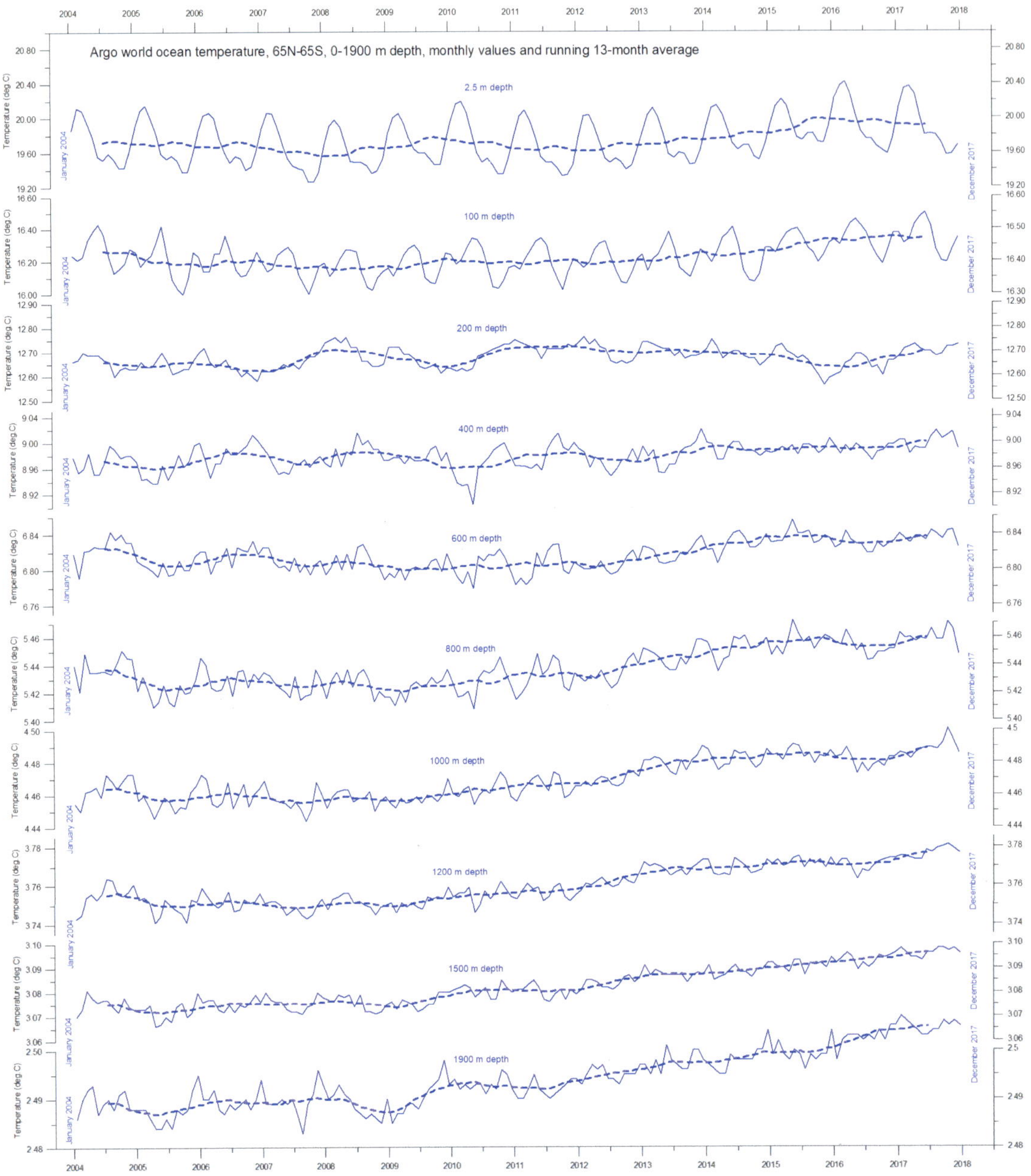

Figure 22: Global ocean temperatures at different depths between 65°N and 65°S. Argo data. The thin line shows monthly values and the thick line shows the running 13-month average. Source: Global Marine Argo Atlas.

For water depths larger than 400 m, however, temperatures are again seen to be increasing. The diagram suggests that this increase first commenced at 1900 m depth around year 2009, and from there gradually has been spreading upwards. At 600 m depth, the present temperature increase began around 2012, that is, about three years later than at 1900 m depth. The timing of these dynamic changes shows that average temperatures in the upper 1900 m of the oceans are influenced by processes operating at greater depths than 1900 m.

Thus, part of the present ocean warming appears to be due to circulation features operating at greater depths than 1900 m, not related directly to processes operating at or near the surface.

This development is also displayed in Figure 23, showing the net change of ocean temperature at different depths, calculated as the net difference between the 12-month averages of 2004 and 2017.

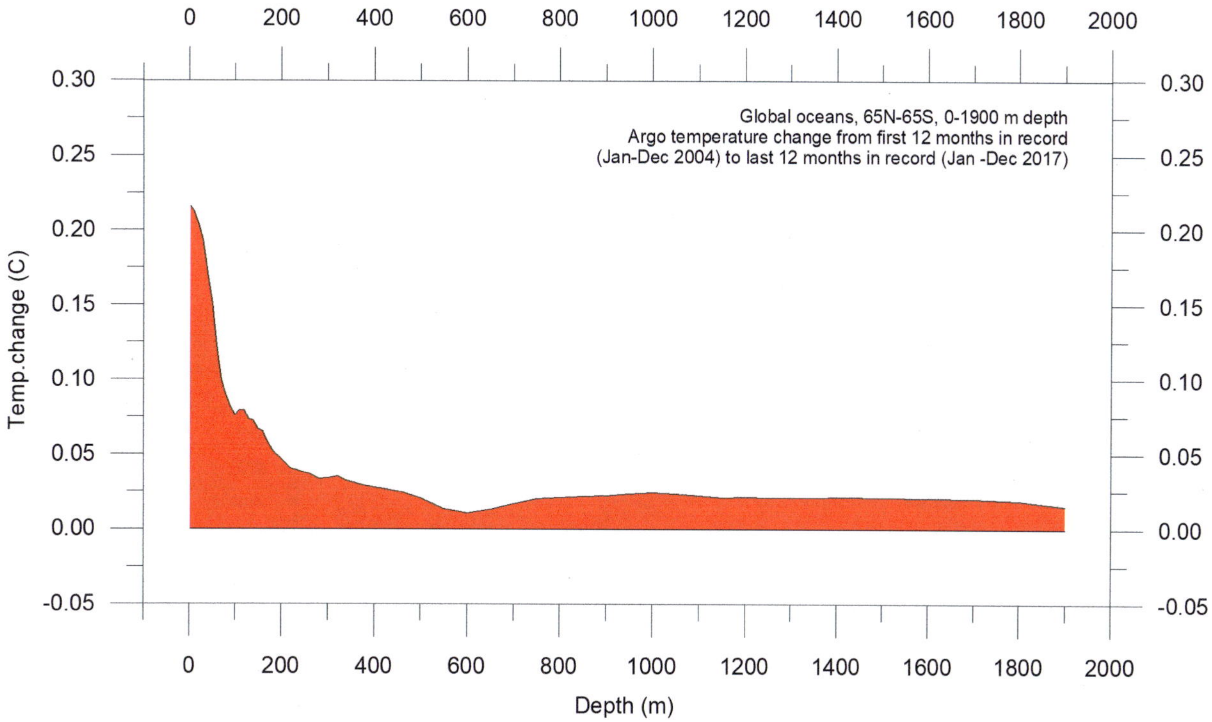

Figure 23: Net change of ocean temperature at different depths.
Calculated as the net difference between the 12-month averages of 2004 and 2017.

16 Regional ocean temperature changes temperatures 0–1900 m depth

Figure 24 shows the latitudinal variation of oceanic temperature changes January-December 2004 versus January-December 2017 for various depths, calculated as in the previous diagram. The three panels show the change in Arctic Oceans (55–65°N), Equatorial Oceans (30N-30°S), and Antarctic Oceans (55–65°S), respectively.

The maximum surface net warming (down to about 100 m depth) affects Equatorial- and Antarctic Oceans, but not the Arctic Oceans. In fact, net cooling down to 1300 m depth is pronounced for the northern oceans. However, the major part of Earth's land areas is in the Northern Hemisphere, wherefore the surface area (and volume) of 'Arctic' oceans is much smaller than the 'Antarctic' oceans, which again is smaller than the 'Equatorial' oceans.

Nevertheless, the contrast in net temperature change experienced 2004–2017 for the different latitudinal bands is instructive. For the two Polar oceans, the data appears to demonstrate the existence of a bi-polar seesaw, as described by Chylek et al. (2010). It is no less interesting, that the near-surface ocean temperature in the two polar regions contrasts with the overall development of sea ice in the two polar regions (see Section 23).

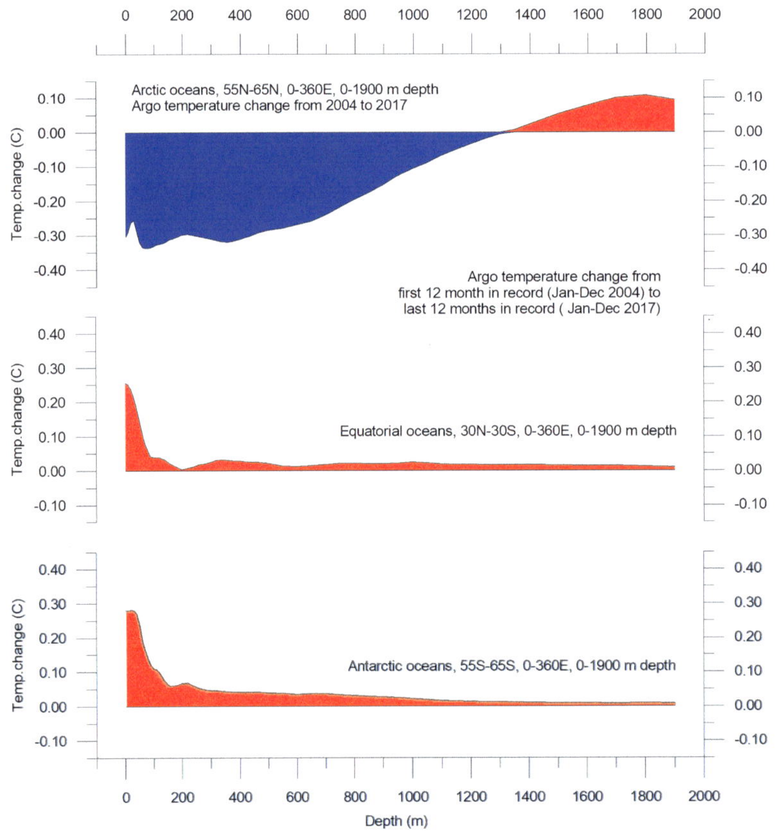

Figure 24: Latitudinal variation of oceanic temperature changes.
January–December 2004 versus January–December 2017 for various depths, calculated as in the previous diagram.

17 Ocean temperature net change 2004–2017 in two north-south sectors

Net changes in temperature along two north-south sectors representing the Atlantic and Pacific oceans (see Figure 25 for locations) are shown in Figure 26 for the Atlantic and Figure 27 for the Pacific . To prepare the diagrams annual average ocean temperatures were compared to annual average temperatures for 2004, representing the first year in the Argo record. To give an insight into ongoing dynamics, the net change in annual average temperatures 2004–2016 is also shown in part (a) of each figure, and the net change for 2004–2017 in part (b). Warm colours indicate net warming since 2004 and blue colours cooling. Due to the spherical shape of the Earth, northern and southern latitudes represent only small ocean volumes, compared to latitudes near the Equator. With this reservation in mind, the diagrams nevertheless reveal interesting features.

The most prominent feature in the Atlantic profile 2004–2017 is a marked cooling north of 30°N, while less pronounced warming dominates south of 20°S, down to depths of 1200 m. The maximum surface warming is confined to the uppermost about 100 m between 45°N and 35°S. At depths greater than 1500 m, warming has also taken place north of 30°N in the 20°W profile. In the South Atlantic, net warming 2004–2017 has taken place between 100 and 1200 m depth, centred around 45°S. The surface warming near Equator and the warming at depth in the South Atlantic is somewhat less pronounced in 2017 than was the case for 2016. Also the net cooling north of 30°N is less pronounced in 2017 than in 2016.

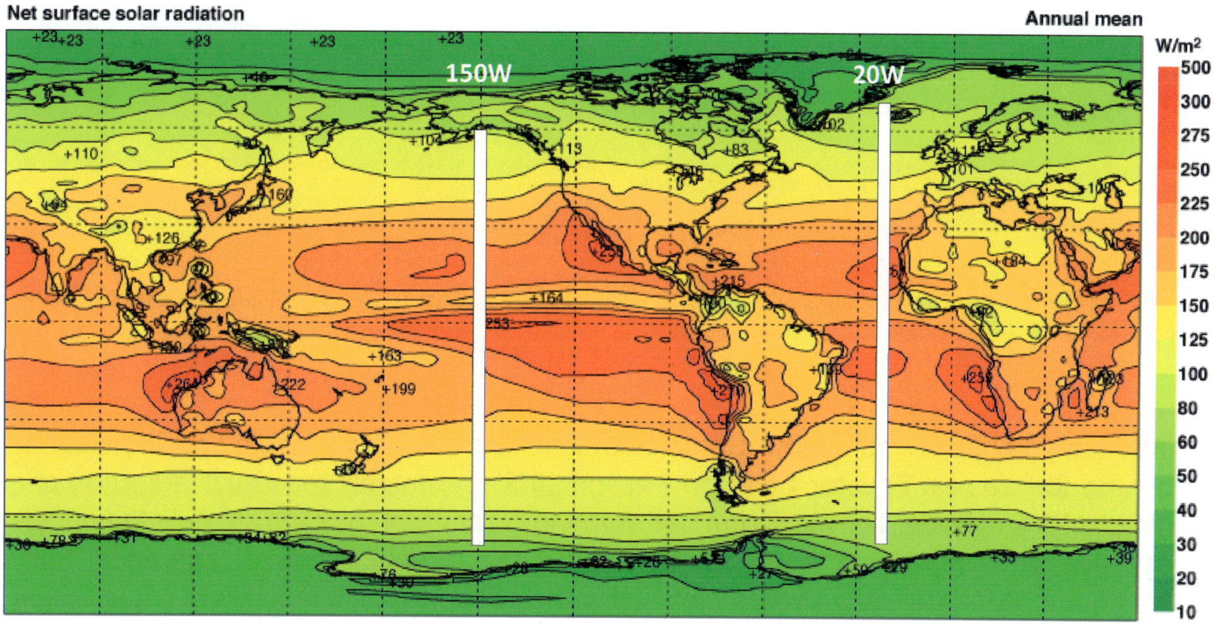

(a)

Figure 25: (a) Map showing average annual mean net surface solar radiation (W/m^2), and the location of the longitudinal profiles discussed in this section.

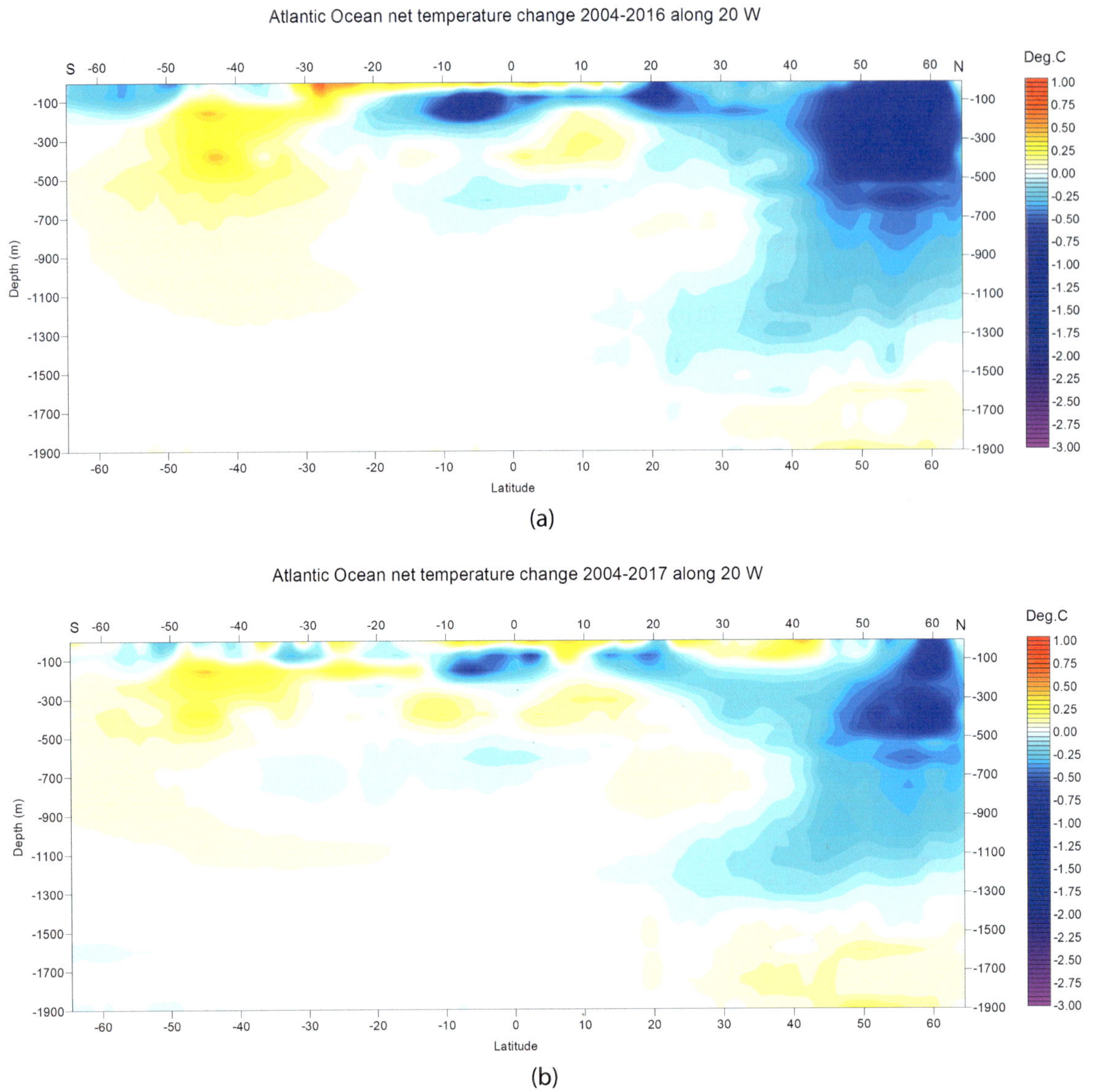

Figure 26: Net changes in temperature along the Atlantic sector for (a) 2004–2016 and (b) 2004–2017.

Along the Pacific sector, the most prominent feature for 2017 is a marked cooling near or below the surface near Equator, between 7°N and 30°S, presumably the visual indication of a coming La Niña episode. The maximum net warming is confined to approximately the uppermost 100 m, with a maximum shortly south of 10°N. The net surface warming around 50–58°N visible on the 2004–2016 diagram has vanished in the 2004–2017 diagram. Slight net warming still characterises water depths of 100–1100 m north of the Equator in 2017. In the Southern Hemisphere south of 45°S, net cooling predominates in nearly all water depths, down to 1900 m. In 2017, marked net cooling since 2004 at 100–600 m depth predominates between 30°S and 25°N, in contrast to the situation in 2016.

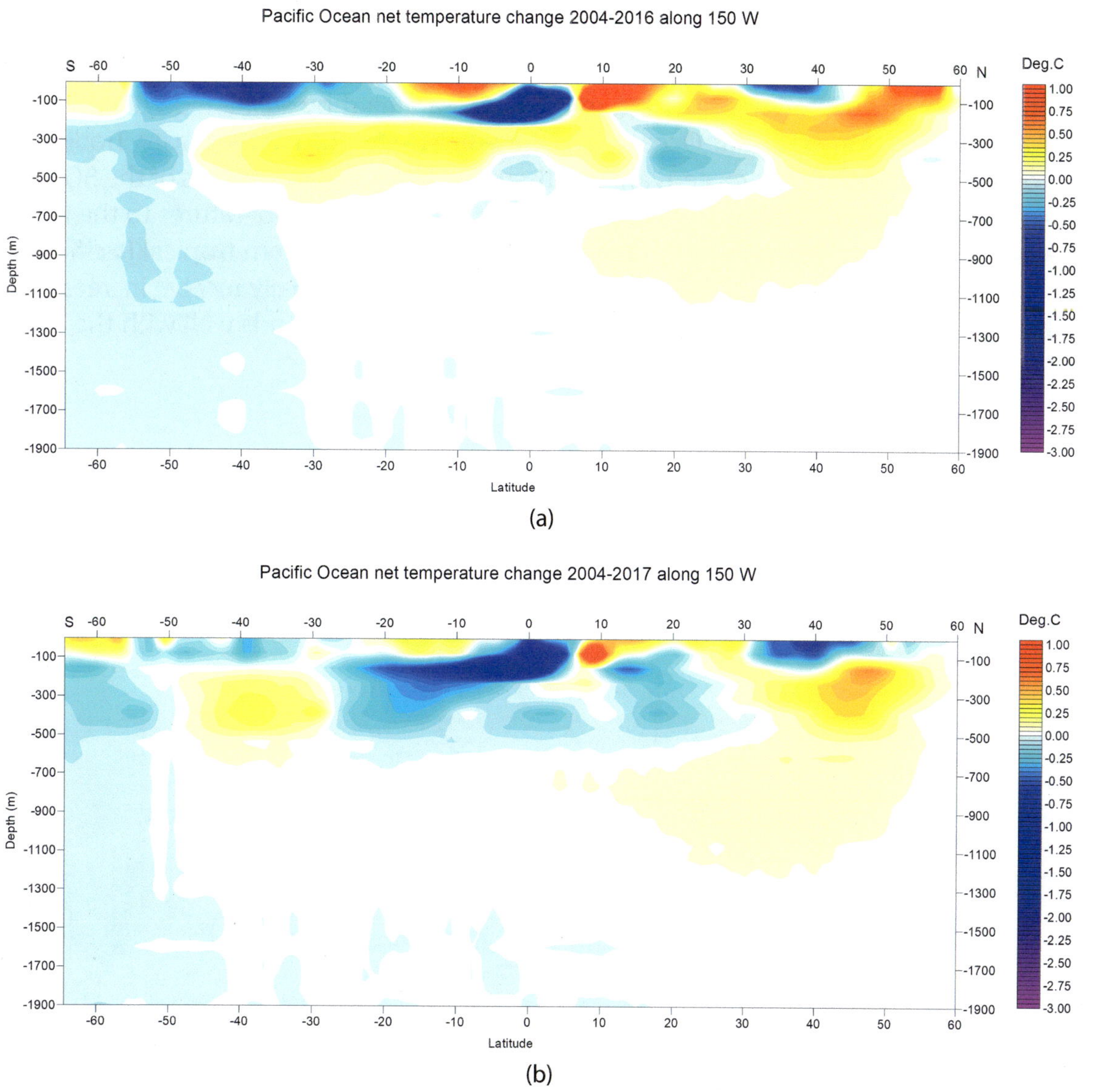

Figure 27: Net changes in temperature along the Pacific sector for (a) 2004–2016 and (b) 2004–2017.

Neither of the diagrams above show to what extent some of the net changes shown are caused by ocean dynamics operating east and west of the two profiles considered; they only display net changes from the beginning to the end of the period considered along the longitudes chosen. For that reason, the diagrams should not be over-interpreted. However, the most prominent surface warming in the two profiles generally appears to coincide with the latitudes of maximum net solar radiation (see Figure 25). In addition, the two profiles suggest an interesting contrast, with the Pacific Ocean mainly warming north of the Equator, and cooling in the south, while the opposite is happening in the Atlantic, with cooling in the north and warming in the south.

18 Southern Oscillation Index

Sustained negative values of the Southern Oscillation Index (SOI; Figure 28) often indicate El Niño episodes. Such negative values are usually accompanied by sustained warming of the central and eastern tropical Pacific Ocean, a decrease in the strength of the Pacific Trade Winds, and a reduction in rainfall over eastern and northern Australia. Positive values of the SOI are associated with stronger Pacific trade winds and higher sea surface temperatures to the north of Australia, indicating La Niña episodes. Waters in the central and eastern tropical Pacific Ocean become cooler during this time. Eastern and northern Australia usually receive increased precipitation during such periods. The recent El Niño episode corresponds well with the low SOI values in 2015–16 (Figure 28).

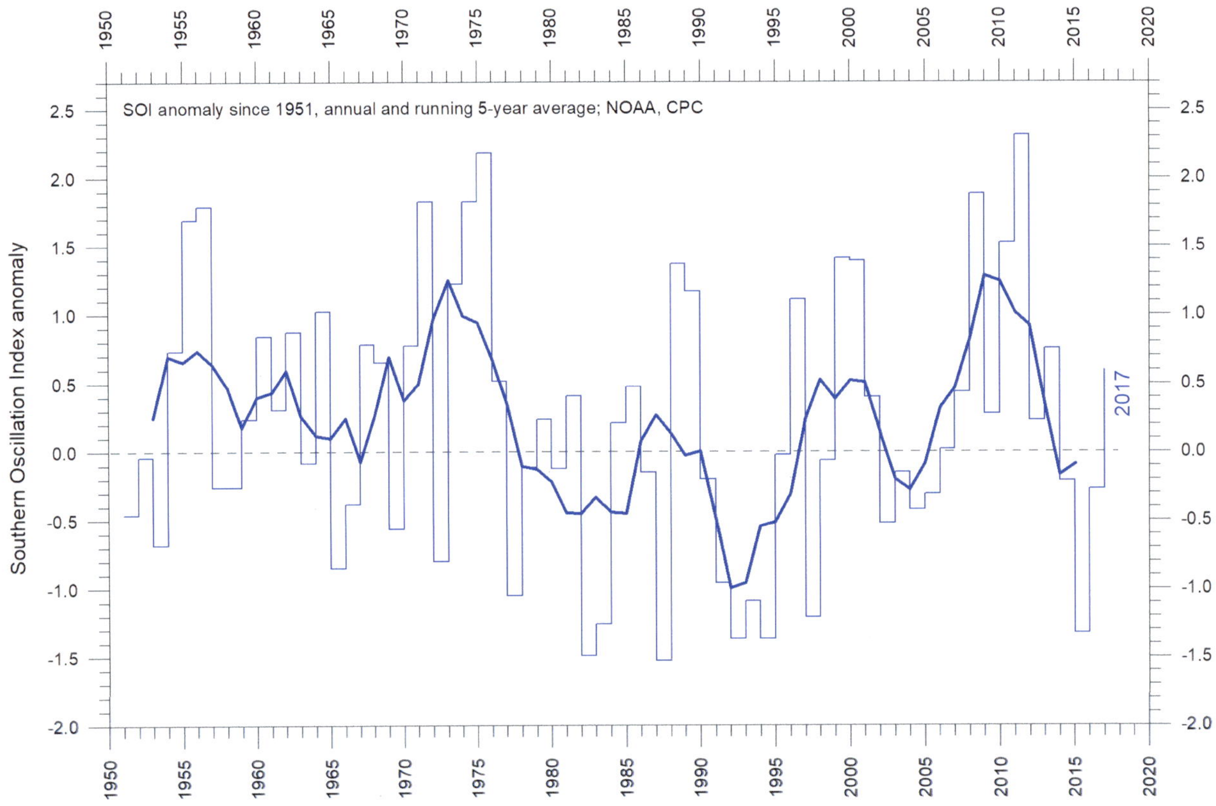

Figure 28: Annual Southern Oscillation Index anomaly since 1951.
According to the NOAA Climate Prediction Center. The SOI is calculated from the monthly or seasonal fluctuations in the air pressure difference between Tahiti and Darwin. The thin line represents annual values, while the thick line is the simple running 5-year average.

19 Pacific Decadal Oscillation

The PDO is a long-lived El Niño-like pattern of Pacific climate variability, with data extending back to January 1900 (Figure 29). Causes for PDO are not currently known, but even in the absence of a theoretical understanding, PDO climate information improves season-to-season and year-to-year climate forecasts for North America because of its strong tendency for multi-season and multi-year persistence. The PDO also appears to be roughly in phase with global temperature changes. Thus, from a societal impacts perspective, recognition of PDO is important because it shows that "normal" climate conditions can vary over time periods comparable to the length of a human's lifetime.

A Fourier frequency analysis (not shown here) shows the PDO record to be influenced by a 5.7-year cycle, and possibly also by a longer about 53-year long cycle. As an example, one marked PDO peak occured shortly before 1940, and another around 1990 (Figure 29).

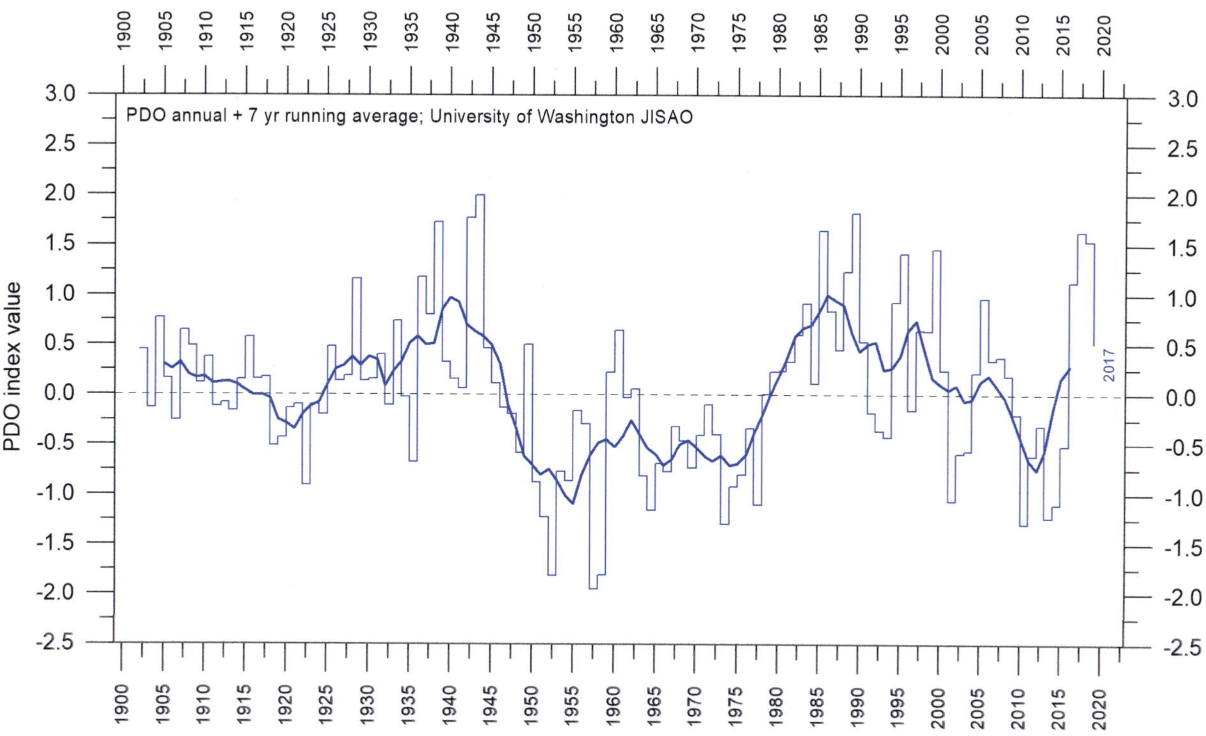

Figure 29: Annual values of the Pacific Decadal Oscillation.
According to the Joint Institute for the Study of the Atmosphere and Ocean, a cooperative Institute between the NOAA and the University of Washington, USA. The thin line shows the annual PDO values, and the thick line is the simple running seven-year average.

20 Atlantic Multidecadal Oscillation

The Atlantic Multidecadal Oscillation (AMO; Figure 30) is a mode of variability occurring in the North Atlantic Ocean sea surface temperature field. The AMO is, in essence, an index of North Atlantic sea surface temperatures (SST).

The AMO index appears to be correlated to air temperatures and rainfall over much of the Northern Hemisphere. The association appears to be high for north-eastern Brazil and African Sahel rainfall and North American and European summer climate. The AMO index also appears to be associated with changes in the frequency of North American droughts and is reflected in the frequency of severe Atlantic hurricanes.

As one example, the AMO index may be related to the past occurrence of major droughts in the US midwest and the southwest. When the AMO is high, these droughts tend to be more frequent or prolonged, and vice-versa for low values of AMO. Two of the most severe droughts of the 20th century in US occurred during the peak AMO values between 1925 and 1965: The Dust Bowl of the 1930s and the 1950s droughts. On the other hand, Florida and the Pacific Northwest tend to be the opposite; a high AMO is here associated with relatively high precipitation.

A Fourier-analysis (not shown here) show the AMO record to be controlled by an about 67-year long cycle, and to a lesser degree by a 3.5-year cycle. Major AMO peaks occurred in around 1875, 1942 and possibly also around 2010 (Figure 30).

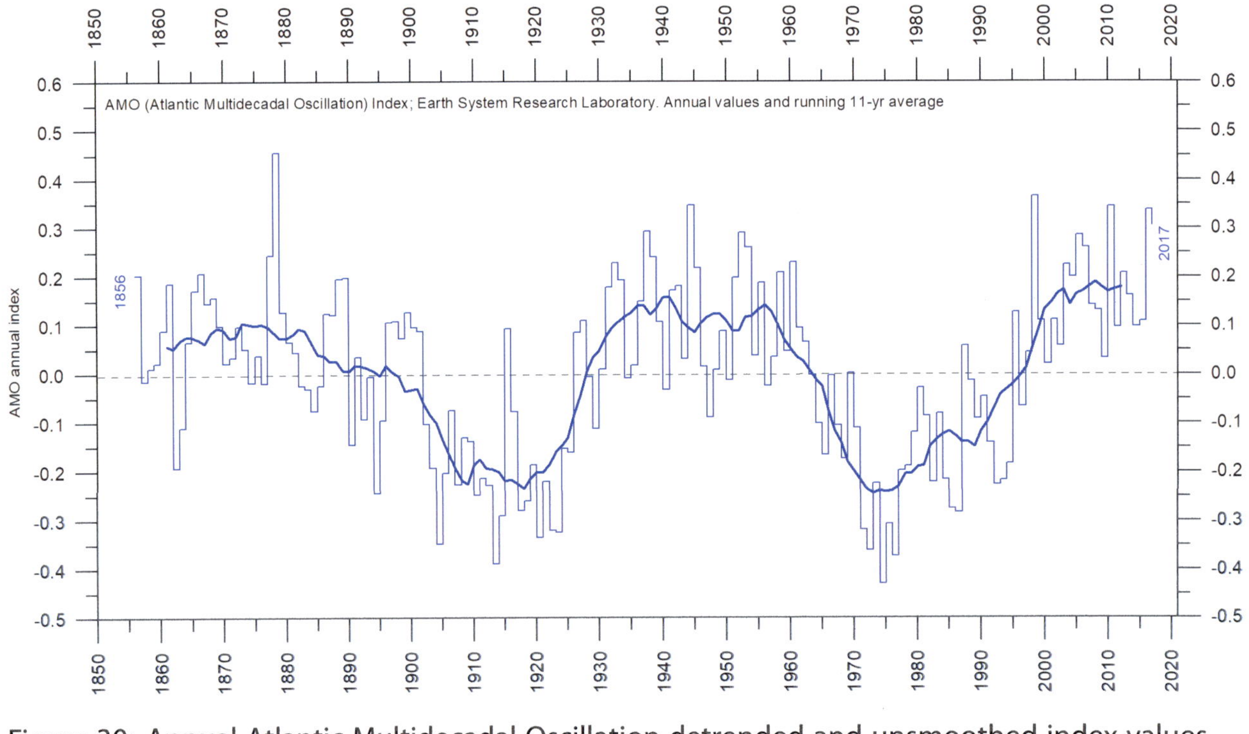

Figure 30: Annual Atlantic Multidecadal Oscillation detrended and unsmoothed index values since 1856.

The thin blue line shows annual values, and the thick line is the simple running 11-year average. Data source: Earth System Research Laboratory, NOAA, USA.

21 Sea-level from satellite altimetry

Satellite altimetry is a new and valuable type of measurement, providing unique insights into the detailed surface topography of the oceans, and their changes. However, it is not a precise tool for estimating changes in global sea level because of several assumptions made when interpreting the original data.

One of the assumptions made during the interpretation of satellite altimetry data is the amount of correction made locally and regionally for glacial isostatic adjustment (GIA). GIA relates to a large-scale, long-term mass transfer from the oceans to the land, in the form of a rhythmic waxing and waning of the large Quaternary ice sheets in North America and northern Europe. This enormous mass transfer causes rhythmic changes in surface load, resulting in viscoelastic mantle flow and elastic effects in the upper crust. No single technique or observational network can give sufficient information on all of the aspects and consequences of GIA, so the assumptions adopted for the interpretation of satellite altimetry data are difficult to verify. The GIA correction introduced in the interpretation of data from satellite altimetry depends upon the type of deglaciation model (for the last glaciation) and upon the type of crust-mantle model that is assumed. Because of this (and additional factors), interpretations of modern global sea level change based on satellite altimetry vary from about 1.7 mm/year to about 3.2 mm/year. In Figure 31 a clear annual variation is apparent, but longer variations may also be in operation (lower panel in figure).

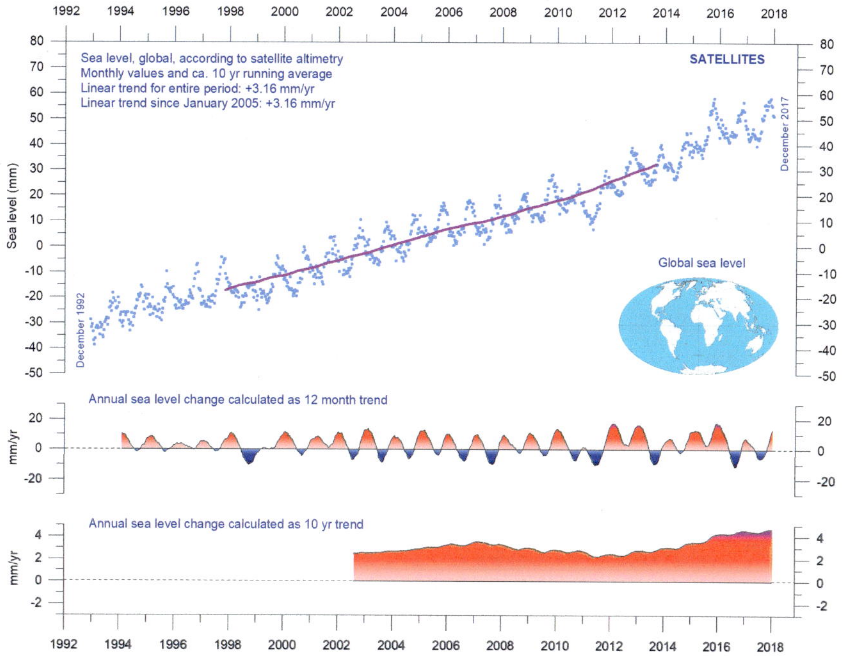

Figure 31: Global sea level since December 1992.

Data according to the Colorado Center for Astrodynamics Research at University of Colorado at Boulder. The blue dots are the individual observations, and the purple line represents the running 121-month (ca. 10-year) average. The two lower panels show the annual sea level change, calculated for 1 and 10-year time windows, respectively. These values are plotted at the end of the interval considered.

22 Sea level from tide-gauges

Tide-gauges are located directly at coastal sites and record the net movement of the local ocean surface in relation to land. Measurements of local relative sea-level change is vital information for coastal planning, and tide-gauge data are therefore directly applicable for planning purposes for coastal installations, in contrast to satellite altimetry.

In a scientific context, the measured net movement of the local sea-level is composed of two local components: the vertical change of the ocean surface and the vertical change of the land surface. For example, a tide-gauge may record an apparent sea-level increase of 3 mm/year. If geodetic measurements show the land to be sinking by 2 mm/year, the real sea level rise is only 1 mm/year (3 minus 2 mm/year). In a global sea-level change context, the value of 1 mm/year is relevant, but in a local coastal planning context the 3 mm/year value obtained from traditional tide gauges is far more relevant. To construct time series of sea-level measurements at each tide gauge, the monthly and annual means must be reduced to a common datum. This reduction is performed by the Permanent Service for Mean Sea Level (PSMSL) making use of the tide gauge datum history provided by the supplying national authority. The Revised Local Reference datum at each station is defined to be approximately 7000 mm below mean sea level, with this arbitrary choice made many years ago to avoid negative numbers in the resulting RLR monthly and annual mean values.

Most tide-gauges are located at sites exposed to tectonic uplift or sinking (the vertical change of the land surface). This widespread vertical instability has several causes, but of course affects the interpretation of data from the individual tide-gauges, although much effort is put into correcting for local tectonic movements.

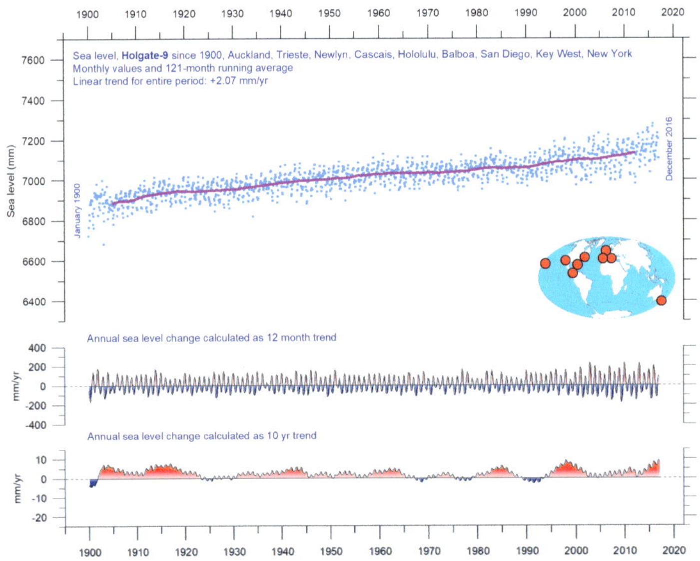

Figure 32: Holgate-9 monthly tide gauge data from PSMSL Data Explorer.

The blue dots are the individual average monthly observations, and the purple line represents the running 121-month (ca. 10-year) average. The two lower panels show the annual sea level change, calculated for 1 and 10-year time windows, respectively. These values are plotted at the end of the interval considered.

Holgate (2007) suggested that the nine stations in Figure 32 captured the variability found in a larger number of stations over the last half century studied previously. For that reason, average values of the Holgate-9 group are interesting to monitor over time. Since 1900, the overall development is nearly linear, with about 2 mm sea level increase indicated (Figure 32).

Of special interest concerning the real short- and long-term sea level change is data obtained by tide-gauges located at tectonic stable sites. One example of such a long record is shown in the diagram below (Korsør, Denmark, since January 1897). This record indicates a stable sea level rise of about 0.84 mm per year, without indication of recent acceleration.

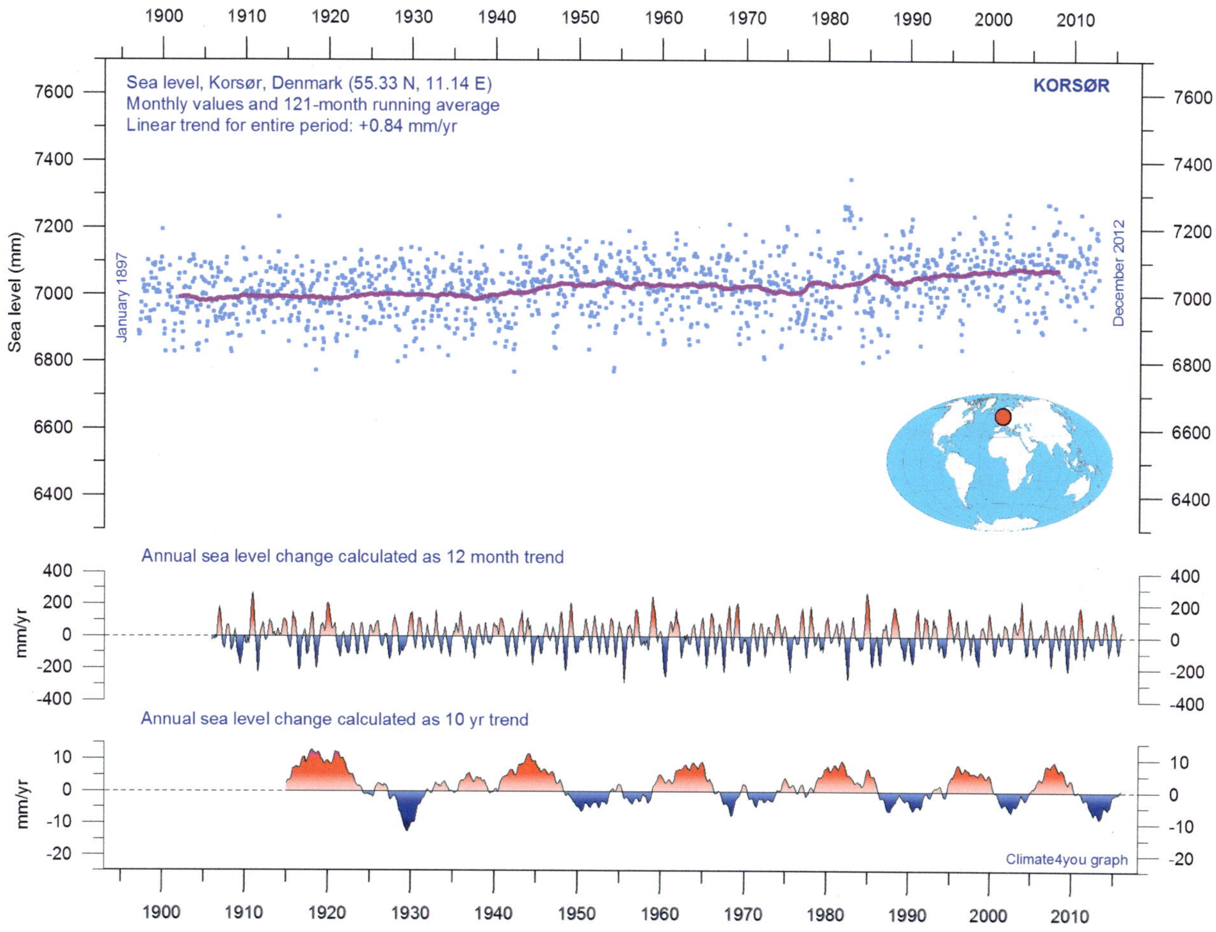

Figure 33: Korsør (Denmark) monthly tide gauge data from PSMSL Data Explorer.
The blue dots are the individual monthly observations, and the purple line represents the running 121-month (ca. 10-year) average. The two lower panels show the annual sea level change, calculated for 1 and 10-year time windows, respectively. These values are plotted at the end of the interval considered.

Data from tide-gauges all over the world suggest an average global sea-level rise of 1–1.5 mm/year, while the record derived from satellites (Section 21) suggests a rise of more than 3 mm/year. The noticeable difference between the two data sets has still no broadly accepted explanation.

23 Global, Arctic and Antarctic sea ice extent

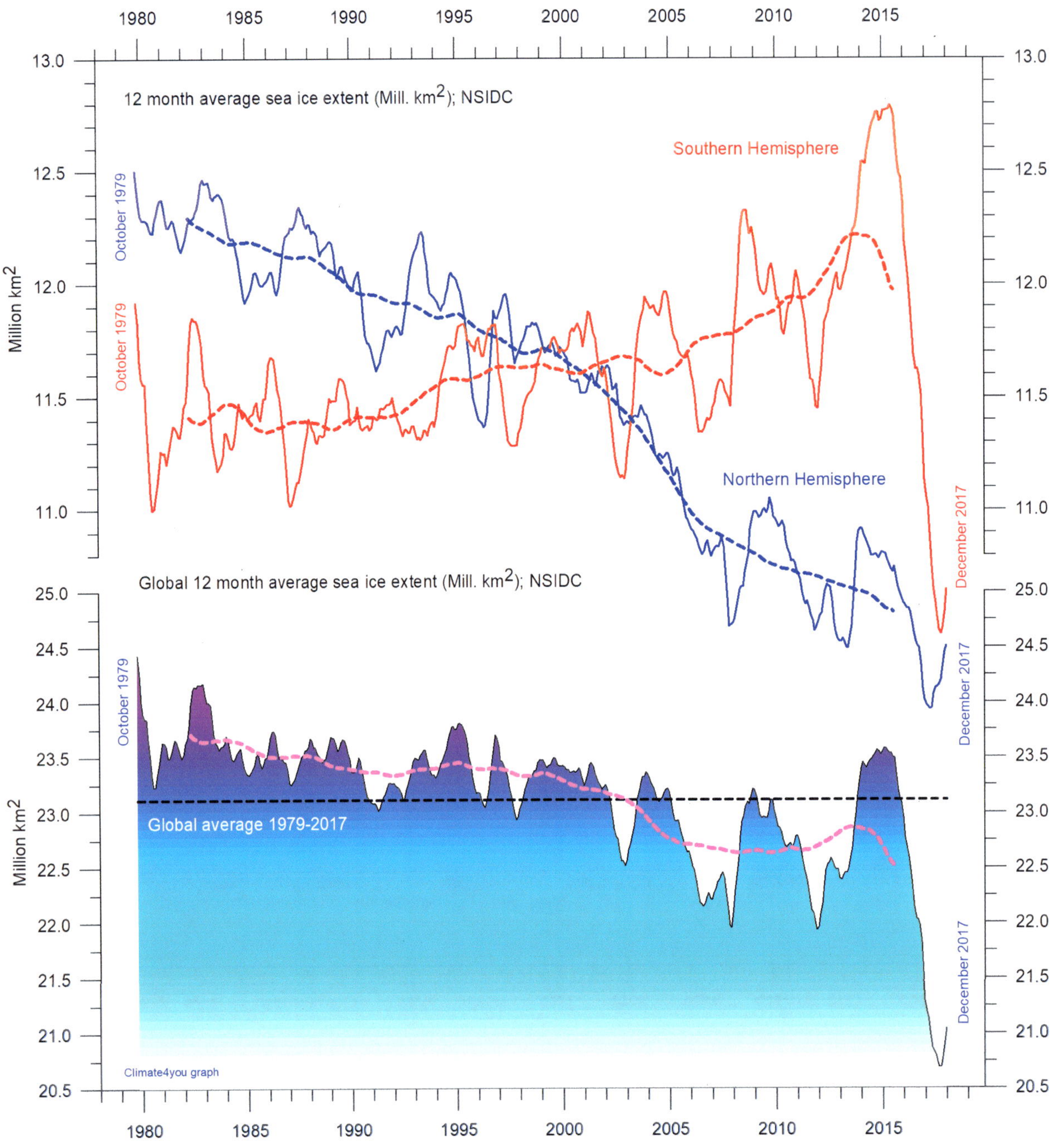

Figure 34: Global and hemispheric 12-month running average sea ice extent since 1979, the satellite-era.

The October 1979 value represents the monthly average of November 1978–October 1979, the November 1979 value represents the average of December 1978–November 1979, etc. The stippled lines represent a 61-month (ca. 5 years) average. Last month included in the 12-month calculations is shown to the right in the diagram. Data source: NSIDC.

Figure 34 shows 12-month average sea ice extents for the Arctic and Antarctic. The trends are in opposite directions. The modern Northern Hemisphere trend towards smaller sea ice extent is clearly displayed by the blue graph, and so is the trend towards a simultaneous increase of Southern Hemisphere sea ice extent.

Both 12-month average graphs are visually characterised by repeated variations, superimposed on the overall trends. This shorter variation is for the Arctic strongly influenced by a 5.3-year periodic variation, while for the Antarctic sea ice a periodic variation of about 4.5 year is important. Both these variations reached their minima simultaneously in 2016, which explains the recent global minima in global sea ice extent.

During the coming years these natural variations may again induce an increase in sea ice extent at both poles, with a derived increase in the 12-month average global sea ice extent as likely result. In fact, this development already appears to have begun by the end of 2017 (Figure 34). However, in the coming years minima and maxima for these variations will not occur synchronously because of their different length, and global minima (or maxima) may be less pronounced than in 2016.

Figure 35 illustrates the overall development of the Arctic sea ice from the end of 2016 to the end of 2017, as published by the Danish Meteorological Institute (DMI). The most conspicuous change has been an overall increase in the estimated ice thickness between the New Siberian Islands (Novosibirskiye Ostrova) and the North Pole. At the same time, the area of thick ice present in late 2016 north of Greenland and Ellesmere Island has been reduced. Presumably, it has at least partly drifted south along the east coast of Greenland.

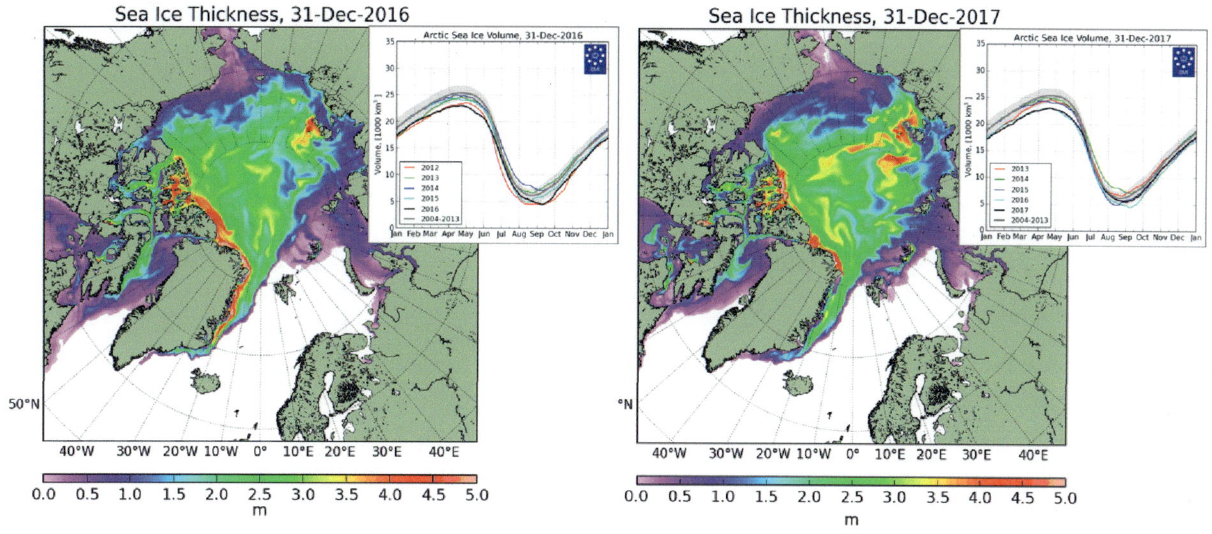

Figure 35: Diagrams showing Arctic sea ice extent and thickness 31 January 2016 (left) and 2017 (right) and the seasonal cycles of the calculated total arctic sea ice volume.

Data according to DMI. The mean sea ice volume and standard deviation for the period 2004–2013 are shown by grey shading in the insert diagrams.

24 Northern Hemisphere snow cover extent

Variations in the global snow cover extent are mainly caused by changes in the Northern Hemisphere, where all the major land areas are located. The Northern Hemisphere snow cover is mainly found north of 30 N (Figure 36). The Southern Hemisphere snow cover extent is essentially controlled by the Antarctic Ice Sheet, and therefore relatively stable.

The Northern Hemisphere snow cover extent is exposed to large local and regional variations from year to year. However, the overall tendency (since 1972) is towards stable Northern-Hemisphere snow conditions (Figure 37).

During the Northern Hemisphere summer, the snow cover usually shrinks to about 2.4 million km^2 (basically controlled by the size of the Greenland Ice Sheet), and during the Northern Hemisphere winter the snow-covered area increases to about 50 million km^2, representing about 33% of planet Earth's total land area.

Considering seasonal changes, the Northern Hemisphere snow cover extent during fall is slightly increasing, the mid-winter extent is basically stable, and the spring extent is slightly declining. In 2017, the Northern Hemisphere snow cover extent was somewhat higher than in 2016, especially during winter, spring and summer (Figure 38).

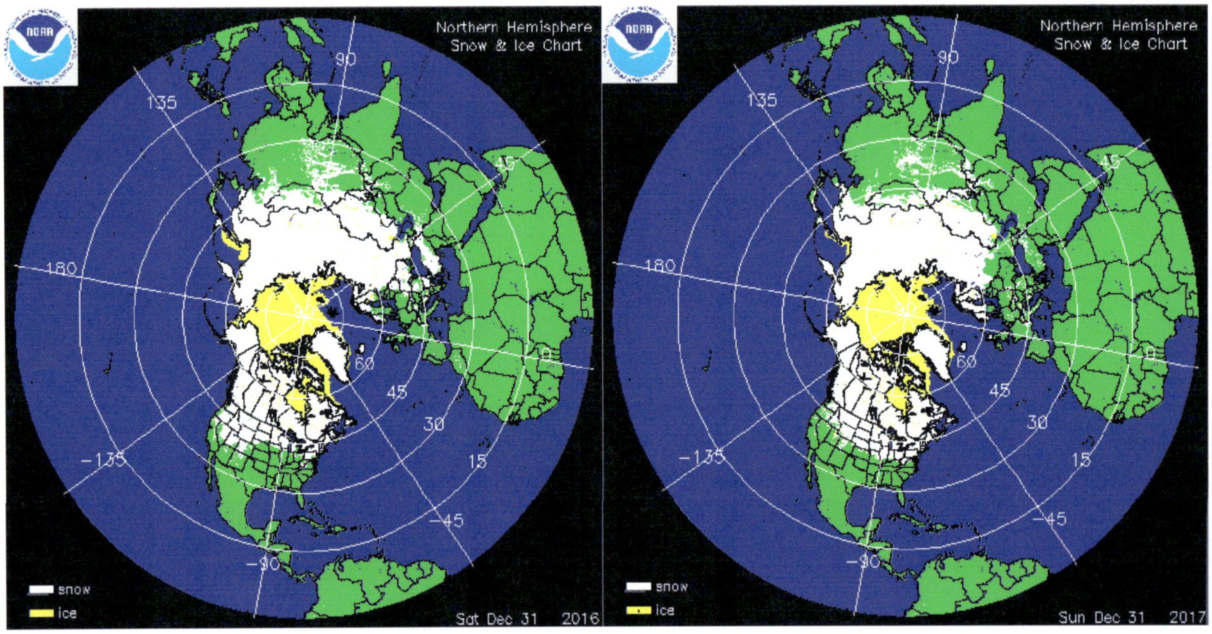

Figure 36: Northern hemisphere snow cover (white) and sea ice (yellow) 31 December 2016 (left) and 2017 (right).

Map source: National Ice Center (NIC).

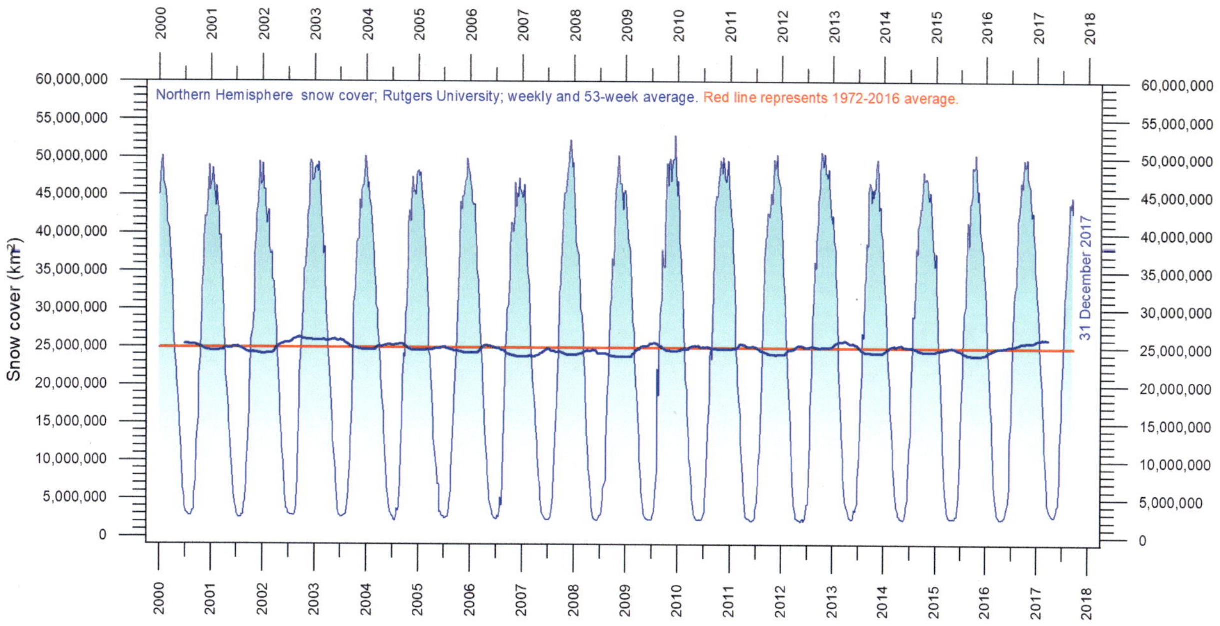

(a)

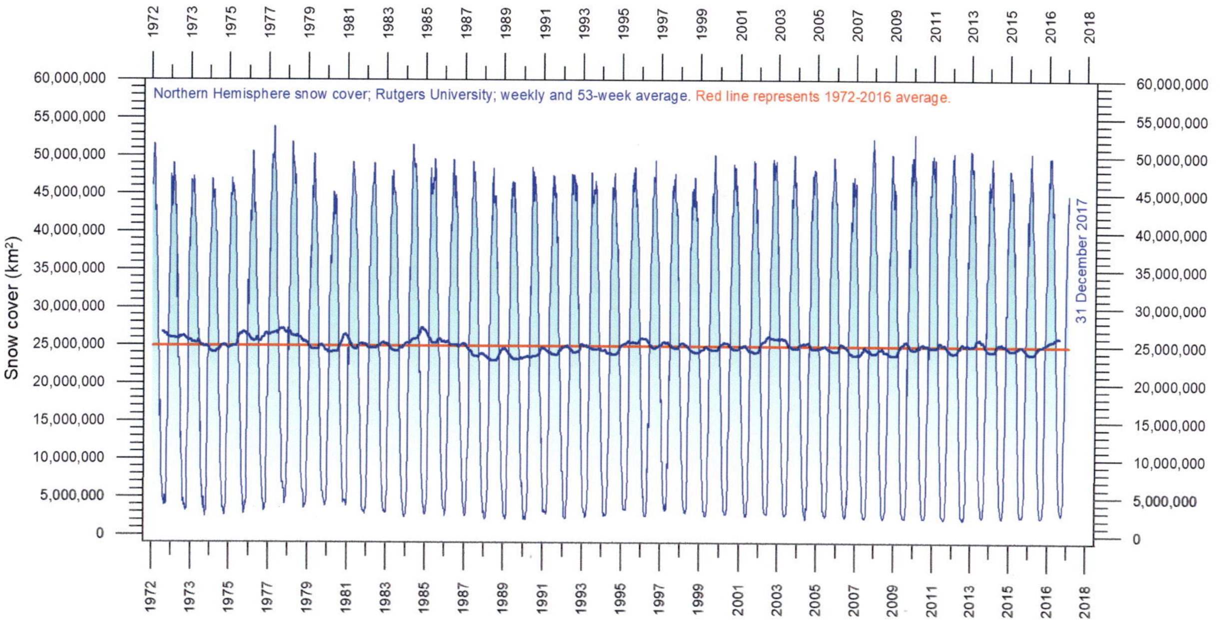

(b)

Figure 37: Northern hemisphere weekly snow cover extent since (a) 2000 and (b) 1972.
Data according to Rutgers University Global Snow Laboratory. The thin blue line is the weekly data, and the thick blue line is the running 53-week average (approximately 1 year). The horizontal red line is the 1972–2016 average.

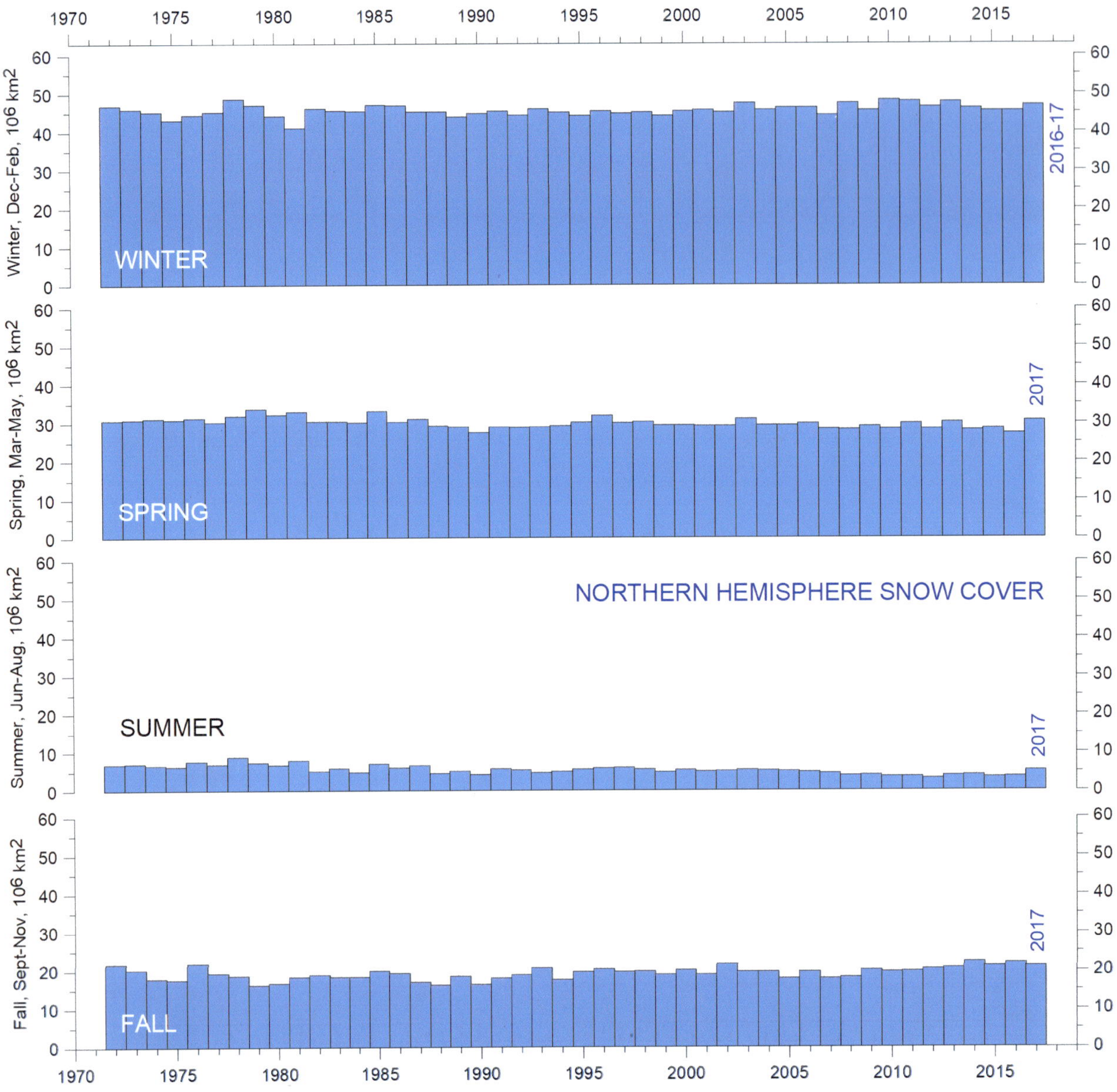

Figure 38: Northern Hemisphere seasonal snow cover since 1972.
Data according to Rutgers University Global Snow Laboratory.

25 Tropical storm and hurricane accumulated cyclone energy (ACE)

Accumulated cyclone energy (ACE) is a measure used to express the energy of individual tropical cyclones and entire tropical cyclone seasons.* The ACE of a season is the sum of the ACE for each storm and considers the number, strength, and duration of all the tropical storms in the season.

The damage potential of a hurricane is proportional to the square or cube of the maximum wind speed, and thus ACE is therefore not only a measure of tropical cyclone activity, but also a measure of the damage potential of an individual cyclone or a season. Existing records (Figures 39 and 40) do not suggest any abnormal cyclone activity in modern years.

The global ACE data (Figure 39) display a variable pattern over time (diagram above), but without any clear trend, as is the case for the Northern- and Southern Hemispheres (Figure 40). The period 1992–1998 was characterised by high values, other peaks were seen 2004–2005 and in 2016, while the periods 1973–1990 and 2002–2015 were characterised by low values. The peaks in 1998 and 2016 coincide with strong El Niño events in the Pacific Ocean (see Section 11). The ACE data and ongoing cyclone dynamics are detailed in Maue (2011).

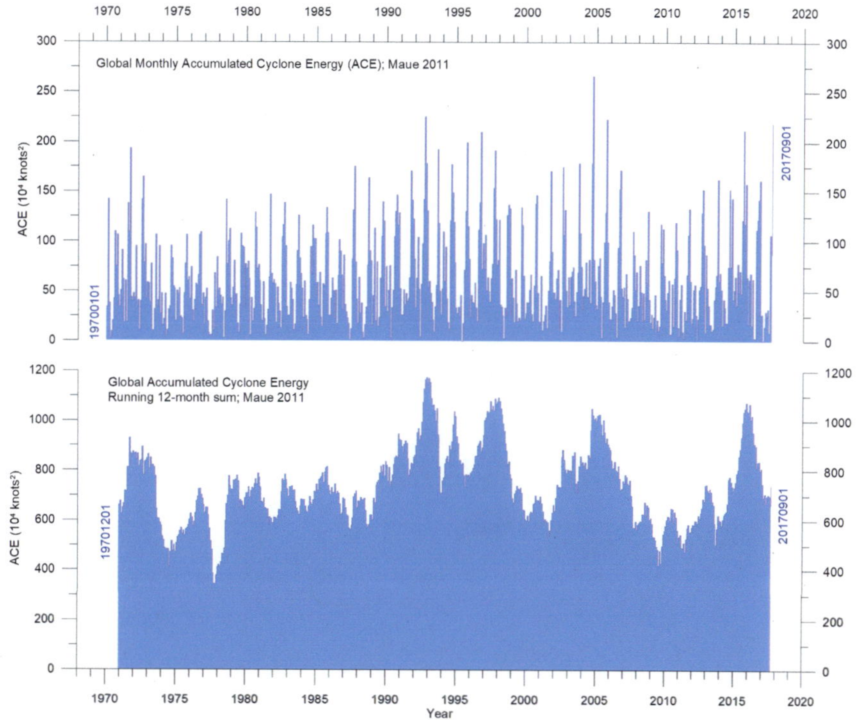

Figure 39: Global tropical storm and hurricane ACE.

Monthly and running 12-month sum, since January 1970. The running 12-month sum (lower panel) is plotted at the end of the time interval considered. Data source: http://wx.graphics/tropical/global_ace.dat

* ACE is calculated as the square of the wind speed every six hours and is then, for usability, is scaled by a factor of 10,000 . The units are 10^4 knots^2.

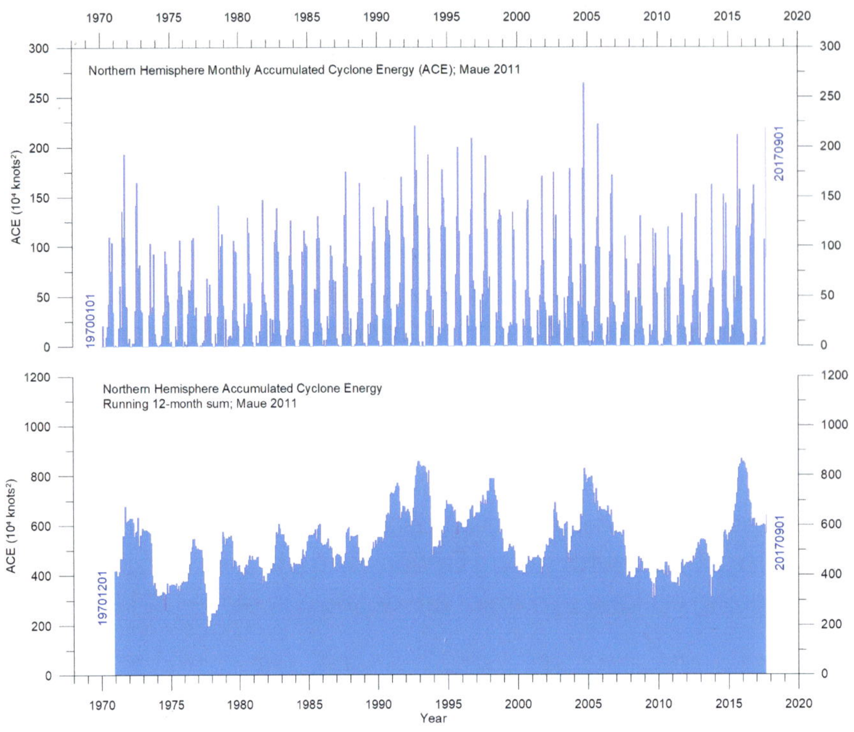

(a)

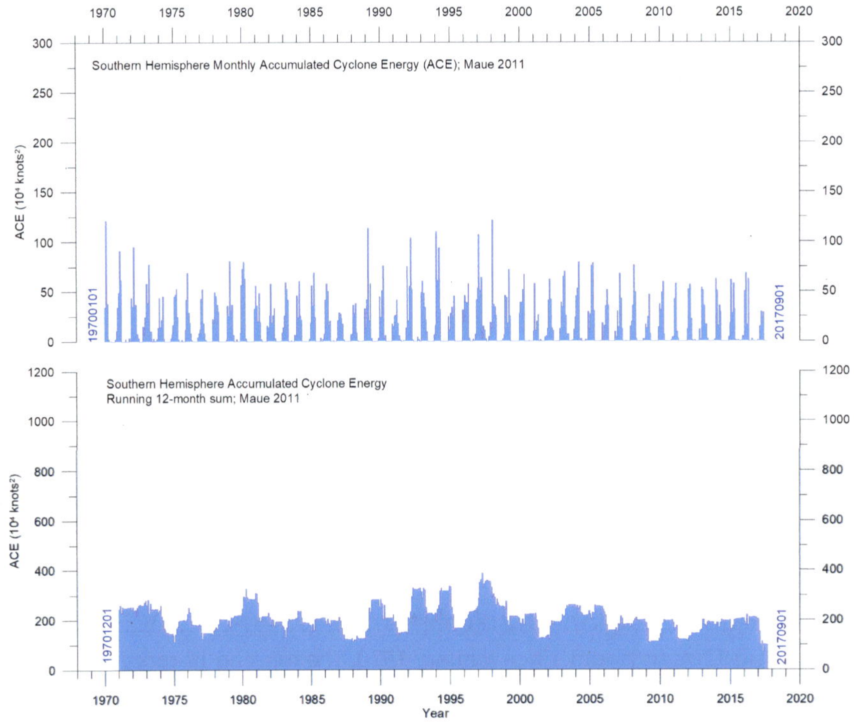

(b)

Figure 40: The Maue (2011) ACE data series: (a) Northern and (b) Southern Hemisphere. Monthly and 12-month sums. The running 12-month sum (lower panels) are plotted at the end of the time intervals considered. Data source: http://wx.graphics/tropical/global_ace.dat

The Northern Hemisphere ACE values (Figure 40a) dominates the global signal (Figure 39) and therefore show similar peaks and lows as displayed by the global data, without any clear trend for the entire observational period. The Northern Hemisphere main cyclone season is June–November. The Southern Hemisphere ACE values (Figure 40b) are lower than for the Northern Hemisphere, and the main cyclone season is December–April.

The Atlantic Oceanographic and Meteorological Laboratory ACE data series goes back to 1850. A Fourier analysis for the Atlantic Basin ((Figure 41)) show the ACE series to be strongly influenced by a periodic variation of about 60 years' duration. At present, since 2002, the Atlantic ACE series is displaying an overall declining trend, but with large interannual variations. The ACE data shown here have not yet been updated beyond 2016, but 2017 turned out to be a very active season in the Atlantic Basin, with substantial damage in the USA and the Caribbean islands (Johnstone and Curry, 2017). The North Atlantic hurricane season often shows above-average activity when La Niña conditions are present in the Pacific during late summer (August–October), as was the case in 2017 (Johnstone and Curry, 2017).

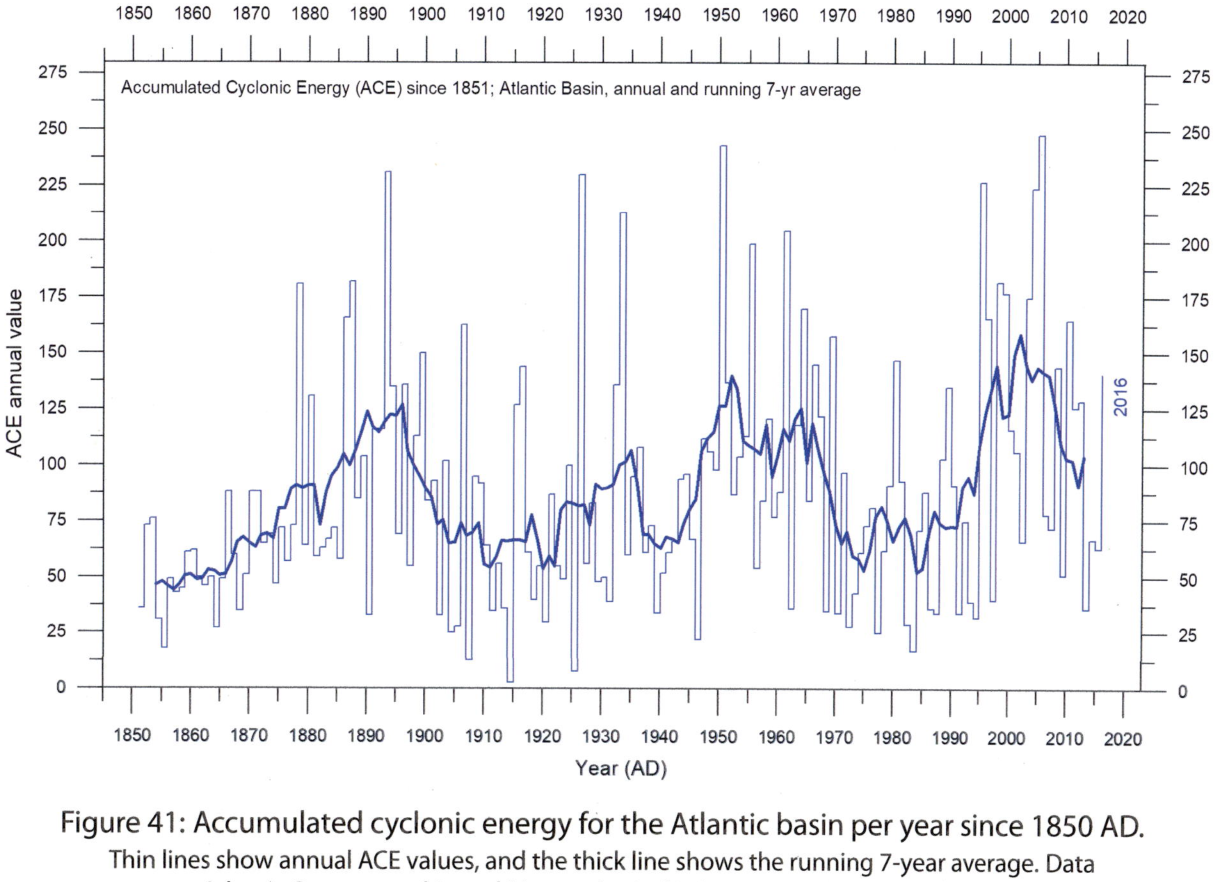

Figure 41: Accumulated cyclonic energy for the Atlantic basin per year since 1850 AD. Thin lines show annual ACE values, and the thick line shows the running 7-year average. Data source: Atlantic Oceanographic and Meteorological Laboratory (AOML), Hurricane research Division.

26 Written references

Chylek, P., Folland, C. K., Lesins, G., and Dubey, M. K. 2010. Twentieth century bipolar seesaw of the Arctic and Antarctic surface air temperatures. *Geophysical Research Letters*, 37, L08703, doi:10.1029/2010GL042793

Holgate, S.J. 2007. On the decadal rates of sea level change during the twentieth century. *Geophysical Research Letters*, 34, L01602, doi:10.1029/2006GL028492

Johnstone, J. and Curry, J. 2017. *Causes and Predictability of the Exceptionally Active 2017 Atlantic Hurricane Season*. Climate Forecast Applications Network. https://curryja.files.wordpress.com/2017/11/hurricane_review_2017-final.pdf.

Maue, R.L. 2011. Recent historically low global tropical cyclone activity. *Geophysical Research Letters*, Vol. 38, L14803, doi:10.1029/2011GL047711

Roemmich, D. and J. Gilson, 2009. The 2004–2008 mean and annual cycle of temperature, salinity, and steric height in the global ocean from the Argo Program. *Progress in Oceanography*, 82, 81–100.

27 Links to data sources (accessed January-February 2018):

AMO, Earth System Research Laboratory, NOAA, USA: s://www.esrl.noaa.gov/psd/data/timeseries/AMO/ Atlantic Oceanographic and Meteorological Laboratory, Hurricane research Division: http://www.aoml.noaa.gov/hrd/tcfaq/E11.html

Colorado Center for Astrodynamics Research: http://sealevel.colorado.edu/

Danish Meteorological Institute (DMI): http://ocean.dmi.dk/arctic/icethickness/thk.uk.php

Earth System Research Laboratory (ESRL): s://www.esrl.noaa.gov/psd/map/clim/olr.shtml

GISS temperature data: s://data.giss.nasa.gov/gistemp/

Global Marine Argo Atlas: http://www.argo.ucsd.edu/Marine_Atlas.html

Goddard Institute for Space Studies (GISS): s://www.giss.nasa.gov/

HadCRUT temperature data: http://hadobs.metoffice.com/

Maue ACE data: Data source: http://wx.graphics/tropical/global_ace.dat and http://wx.graphics/tropical/

National Ice Center (NIC). http://www.natice.noaa.gov/pub/ims/ims_gif/DATA/cursnow.gif

National Snow and Ice Data Center (NSIDC): http://nsidc.org/data/seaice_index/index.html

NCDC temperature data: s://www.ncdc.noaa.gov/monitoring-references/faq/

Ocean temperatures from Argo floats: http://www.argo.ucsd.edu/

Oceanic Niño Index (ONI): http://www.cpc.ncep.noaa.gov/products/analysis_monitoring/ensostuff/ensoyears.shtml

Outgoing long wave radiation (OLR): s://www.esrl.noaa.gov/psd/map/clim/olr.shtml

PDO, Joint Institute for the Study of the Atmosphere and Ocean (JISAO): http://research.jisao.washington.edu/pdo/PDO.latest

Permanent Service for Mean Sea Level: http://www.psmsl.org/

PSMSL Data Explorer: http://www.psmsl.org/data/obtaining/map.html

Rutgers University Global Snow Laboratory: http://climate.rutgers.edu/snowcover/index.php

RSS temperature data: http://www.remss.com/measurements/upper-air-temperature

Sea level from satellites: http://sealevel.colorado.edu/files/current/sl_global.txt

Sea level from tide-gauges: http://www.psmsl.org/data/obtaining/map.html

Sea ice extent Danish Meteorological Institute (DMI): http://ocean.dmi.dk/arctic/icethickness/thk.uk.php

Southern Oscillation Index (SOI): http://www.cpc.noaa.gov/data/indices/soi

Tropical storm and hurricane accumulated cyclone energy (ACE): http://wx.graphics/tropical/global_ace.dat

UAH temperature data: http://www.nsstc.uah.edu/data/msu/v6.0/tlt/uahncdc_lt_6.0.txt

About the Global Warming Policy Foundation

The Global Warming Policy Foundation is an all-party and non-party think tank and a registered educational charity which, while openminded on the contested science of global warming, is deeply concerned about the costs and other implications of many of the policies currently being advocated.

Our main focus is to analyse global warming policies and their economic and other implications. Our aim is to provide the most robust and reliable economic analysis and advice. Above all we seek to inform the media, politicians and the public, in a newsworthy way, on the subject in general and on the misinformation to which they are all too frequently being subjected at the present time.

The key to the success of the GWPF is the trust and credibility that we have earned in the eyes of a growing number of policy makers, journalists and the interested public. The GWPF is funded overwhelmingly by voluntary donations from a number of private individuals and charitable trusts. In order to make clear its complete independence, it does not accept gifts from either energy companies or anyone with a significant interest in an energy company.

Views expressed in the publications of the Global Warming Policy Foundation are those of the authors, not those of the GWPF, its trustees, its Academic Advisory Council members or its directors.

Made in United States
Troutdale, OR
11/03/2026

56844698R00036